Microcontroller Based
Applied Digital Control

Microcontroller Based Applied Digital Control

Dogan Ibrahim

Department of Computer Engineering
Near East University, Cyprus

John Wiley & Sons, Ltd

Other Wiley Editorial Offices

John Wiley & Sons Inc., 111 River Street, Hoboken, NJ 07030, USA

Jossey-Bass, 989 Market Street, San Francisco, CA 94103-1741, USA

Wiley-VCH Verlag GmbH, Boschstr. 12, D-69469 Weinheim, Germany

John Wiley & Sons Australia Ltd, 42 McDougall Street, Milton, Queensland 4064, Australia

John Wiley & Sons (Asia) Pte Ltd, 2 Clementi Loop #02-01, Jin Xing Distripark, Singapore 129809

John Wiley & Sons Canada Ltd, 22 Worcester Road, Etobicoke, Ontario, Canada M9W 1L1

Wiley also publishes its books in a variety of electronic formats. Some content that appears in print may not be available in electronic books.

Library of Congress Cataloging-in-Publication Data

Ibrahim, Dogan.
    Microcontroller based applied digital control / Dogan Ibrahim.
    p. cm.
  ISBN 0-470-86335-8
  1. Process control—Data processing.   2. Digital control systems—Design and construction.   3. Microprocessors.
  I. Title.
  TS156.8.I126 2006
  629.8′9—dc22   2005030149

British Library Cataloguing in Publication Data

A catalogue record for this book is available from the British Library

ISBN-13 978-0-470-86335-0 (HB)
ISBN-10 0-470-86335-8 (HB)

Typeset in 10/12pt Times by TechBooks, New Delhi, India

Contents

Preface

Computers now form an integral part of most real-time control systems. With the advent of the microprocessors and microcontrollers in the last few decades the use of computers in control applications has been ever growing. Microcontrollers are single-chip computers which can be used to control real-time systems. Such controllers are also referred to as embedded real-time computers. These devices are low-cost, single-chip and easy to program. Microcontrollers have traditionally been programmed using the assembly language of the target processor. It is now possible to program these devices using high-level languages such as BASIC, PASCAL, or C. As a result of this, very complex control algorithms can be developed and implemented on the microcontrollers.

This book is about the theory and practice of microcontroller based automatic control systems engineering. A previous knowledge of microcontroller hardware or software is not required, but the reader will find it useful to have some knowledge of a computer programming language.

Chapter 1 of the book presents a brief introduction to the control systems and the elements of computer based control systems. Some previous knowledge of the theory of continuous-time control systems is helpful in understanding this material.

Chapter 2 is about system modelling. Modelling a dynamic system is the starting point in control engineering. Models of various mechanical, electrical, and fluid systems are introduced in this chapter.

Chapter 3 is devoted to the popular PIC microcontroller family which is described and used in this book. The PIC family is one of the most widely used microcontrollers in commercial and industrial applications. The chapter describes the features of this family, and basic application notes are also given.

The book is based on the C programming language known as *PICC Lite*. This is distributed free by Hi-Tech Software and is used to program the PIC family of microcontrollers. Chapter 4 gives a brief introduction to the features of this language.

The microcontroller project development cycle is described in some detail in Chapter 5. The knowledge of the microcontroller development cycle is important as the developed controller algorithm has to be implemented on the target microcontroller.

Chapters 6 and 7 are devoted to the analysis of discrete-time systems. The terms *discrete-time system, sampled-data system* and *digital control system* are all used interchangeably in the book and refer to the same topic. The sampling process, z-transforms, and the time response of discrete-time systems are explained in detail in these two chapters.

The stability of a control system is one of the most important topics in control engineering. Chapter 8 analyses the stability of digital control systems with examples, using the various well-established analytical and graphical stability techniques.

The analysis and design of digital controllers are described in Chapter 9, where various digital controller algorithms are developed with examples.

After a digital controller is designed, it has to be implemented on the microcontroller; this is known as the realization of the controller. Chapter 10 describes various realization techniques, describing the advantages and disadvantages of each technique. Programming examples are given to show how a particular realization can be programmed and implemented on a microcontroller.

Finally, Chapter 11 presents a case study. A liquid level control system is modelled and then a suitable digital controller algorithm is developed. The algorithm is then implemented on a PIC microcontroller. The time response of the system is given, along with a full program listing of the algorithm .

Many people have assisted in the production and development of this book. In particular, I wish to acknowledge the contribution of the students and staff members of the Computer Engineering Department of the Near East University.

Dogan Ibrahim
Near East University

1

Introduction

1.1 THE IDEA OF SYSTEM CONTROL

Control engineering is concerned with controlling a dynamic system or plant. A dynamic system can be a mechanical system, an electrical system, a fluid system, a thermal system, or a combination of two or more types of system. The behaviour of a dynamic system is described by differential equations. Given the model (differential equation), the inputs and the initial conditions, we can easily calculate the system output.

A plant can have one or more inputs and one or more outputs. Generally a plant is a continuous-time system where the inputs and outputs are also continuous in time. For example, an electromagnetic motor is a continuous-time plant whose input (current or voltage) and output (rotation) are also continuous signals. A control engineer manipulates the input variables and shapes the response of a plant in an attempt to influence the output variables such that a required response can be obtained.

A plant is an *open-loop* system where inputs are applied to drive the outputs. For example, a voltage is applied to a motor to cause it to rotate. In an open-loop system there is no knowledge of the system output. The motor is expected to rotate when a voltage is applied across its terminals, but we do not know by how much it rotates since there is no knowledge about the output of the system. If the motor shaft is loaded and the motor slows down there is no knowledge about this. A plant may also have disturbances affecting its behaviour and in an open-loop system there is no way to know, or to minimize these disturbances.

Figure 1.1 shows an open-loop system where the system input is expected to drive the system output to a known point (e.g. to rotate the motor shaft at a specified rate). This is a single-input, single-output (SISO) system, since there is only one input and also only one output is available.

In general, systems can have multiple inputs and multiple outputs (MIMO). Because of the unknowns in the system model and the effects of external disturbances the open-loop control is not attractive. There is a better way to control the system, and this is by using a sensor to measure the output and then comparing this output with what we would like to see at the system output. The difference between the desired output value and the actual output value is called the *error signal*. The error signal is used to force the system output to a point such that the desired output value and the actual output value are equal. This is termed *closed-loop* control, or feedback control. Figure 1.2 shows a typical closed-loop system. One of the advantages of closed-loop control is the ability to compensate for disturbances and yield the correct output even in the presence of disturbances. A *controller* (or a *compensator*) is usually employed to read the error signal and drive the plant in such a way that the error tends to zero.

Microcontroller Based Applied Digital Control   D. Ibrahim
© 2006 John Wiley & Sons, Ltd

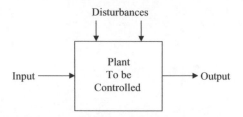

Figure 1.1   Open-loop system

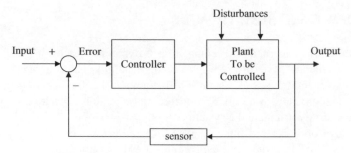

Figure 1.2   Closed-loop system

Closed-loop systems have the advantage of greater accuracy than open-loop systems. They are also less sensitive to disturbances and changes in the environment. The time response and the steady-state error can be controlled in a closed-loop system.

Sensors are devices which measure the plant output. For example, a thermistor is a sensor used to measure the temperature. Similarly, a tachogenerator is a sensor used to measure the rotational speed of a motor, and an accelerometer is used to measure the acceleration of a moving body. Most sensors are analog devices and their outputs are analog signals (e.g. voltage or current). These sensors can be used directly in continuous-time systems. For example, the system shown in Figure 1.2 is a continuous-time system with analog sensors, analog inputs and analog outputs. Analog sensors cannot be connected directly to a digital computer. An analog-to-digital (A/D) converter is needed to convert the analog output into digital form so that the output can be connected to a digital computer. Some sensors (e.g. temperature sensors) provide digital outputs and can be directly connected to a digital computer.

With the advent of the digital computer and low-cost microcontroller processing elements, control engineers began to use these programmable devices in control systems. A digital computer can keep track of the various signals in a system and can make intelligent decisions about the implementation of a control strategy.

1.2   COMPUTER IN THE LOOP

Most control engineering applications nowadays are computer based, where a digital computer or a microcontroller is used as the controller. Figure 1.3 shows a typical computer controlled system. Here, it is assumed that the error signal is analog and an A/D converter is used to convert the signal into digital form so that it can be read by the computer. The A/D converter

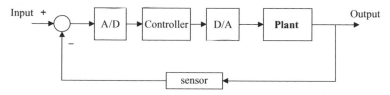

Figure 1.3 Typical digital control system

samples the signal periodically and then converts these samples into a digital word suitable for processing by the digital computer. The computer runs a controller algorithm (a piece of software) to implement the required actions so that the output of the plant responds as desired. The output of a digital computer is a digital signal, and this is normally converted into analog form by using a digital-to-analog (D/A) converter. The operation of a D/A converter is usually approximated by a zero-order hold transfer function.

There are many microcontrollers that incorporate built-in A/D and D/A converter circuits. These microcontrollers can be connected directly to analog signals, and to the plant.

In Figure 1.3 the reference set-point, sensor output, and the plant input and output are all assumed to be analog. Figure 1.4 shows the block diagram of the system in Figure 1.3 where the A/D converter is shown as a sampler. Most modern microcontrollers include built-in A/D and D/A converters, and these have been incorporated into the microcontroller in Figure 1.4.

There are other variations of the basic digital control system. In Figure 1.5 another type of digital control system is shown where the reference set-point is read from the keyboard or is hard-coded into the control algorithm. Since the sensor output is analog, it is converted into digital form using an A/D converter and the resulting digital signal is fed to the computer where the error signal is calculated and is used to implement the control algorithm.

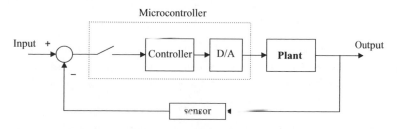

Figure 1.4 Block diagram of a digital control system

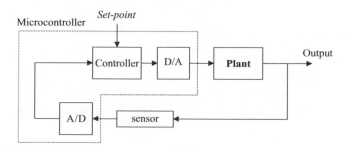

Figure 1.5 Another form of digital control

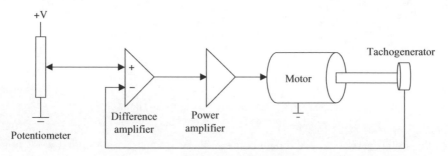

Figure 1.6   Typical analog speed control system

The purpose of developing the digital control theory is to be able to understand, design and build control systems where a computer is used as the controller in the system. In addition to the normal control task, a computer can perform supervisory functions, such as reading data from a keyboard, displaying data on a screen or liquid crystal display, turning a light or a buzzer on or off and so on.

Figure 1.6 shows a typical closed-loop analog speed control system where the desired speed of the motor is set using a potentiometer. A tachogenerator produces a voltage proportional to the speed of the motor, and this signal is used in a feedback loop and is subtracted from the desired value in order to generate the error signal. Based on this error signal the power amplifier drives the motor to obtain the desired speed. The motor will rotate at the desired speed as long as the error signal is zero.

The equivalent digital speed control system is shown in Figure 1.7. Here, the desired speed is entered from the keyboard into the digital controller. The controller also receives the converted output signal of the tachogenerator. The error signal is calculated by the controller by subtracting the tachogenerator reading from the desired speed. A D/A converter is then used to convert the signal into analog form and feed the power amplifier. The power amplifier then drives the motor.

Since the speed control can be achieved by using an analog approach, one is tempted to ask why use digital computers. Digital computers in 1960s were very large and very expensive devices and their use as controllers was not justified. They could only be used in very large

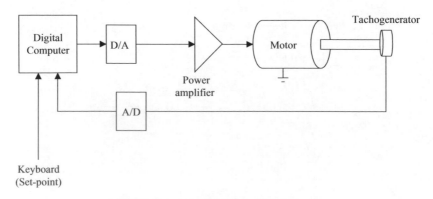

Figure 1.7   Digital speed control system

and expensive plants, such as large chemical processing plants or oil refineries. Since the introduction of microprocessors in the early 1970s the cost and size of digital computers have been greatly reduced. Early microprocessors, such as the Intel 8085 or the Mostek Z80, were very limited and required several chips before they could be used as processing elements. The required chips were read-only memory (ROM) to store the user program, random-access memory (RAM) to store the user data, input–output (I/O) circuitry, A/D and D/A converters, interrupt logic, and timer circuits. By the time all these chips were put together the chip count, power consumption, and complexity of the basic hardware were considerable. These controllers were in the form of microcomputers which could be used in many medium and large digital control applications.

Interest in digital control has grown rapidly in the last several decades since the introduction of *microcontrollers*. A microcontroller is a single-chip computer, including most of a computer's features, but in limited sizes. Today, there are hundreds of different types of microcontrollers, ranging from 8-pin devices to 40-pin, or even 64- or higher pin devices. For example, the PIC16F877 is an 8-bit, 40-pin microcontroller with the following features:

- operation up to 20 MHz;
- 8K flash program memory;
- 368 bytes RAM memory;
- 256 bytes electrically erasable programmable read-only memory (EEPROM) memory;
- 15 types of interrupts;
- 33 bits of parallel I/O capability;
- 2 timers;
- universal synchronous–asynchronous receiver/transmitter (USART) serial communications;
- 10-bit, 8-channel A/D converter;
- 2 analog comparators;
- 33 instructions;
- programming in assembly or high-level languages;
- low cost (approximately $10 each).

Flash memory is nonvolatile and is used to store the user program. This memory can be erased and reprogrammed electrically. EEPROM memory is used to store nonvolatile user data and can be written to or read from under program control. The microcontroller has 8K program memory, which is quite large for control based applications. In addition, the RAM memory is 368 bytes, which again is quite large for control based applications.

1.3 CENTRALIZED AND DISTRIBUTED CONTROL SYSTEMS

Until the beginning of 1980s, computer control was strictly *centralized*. Usually a single large computer or minicomputer (e.g. the DEC PDP11 series) was used to control the plant. The computer, associated power supplies, input–output, keyboard and display unit were all situated in a central location. The advantages of centralized control are as follows:

- It is easy to manage the computer.
- Only one computer is used.
- Less number of people are required.

In a centralized control system, the controller algorithm is implemented in a single central computer. Hence, all sensors, actuators, input units and output units must be connected directly to this central computer.

Today, *distributed control* is more widely used. A distributed control system (DCS) consists of a number of computers installed at different locations, each performing an independent control action. Distributed control has emerged as a result of the sharp decrease in price, and the consequent widespread use, of computers. Also, the development of computer networks has made it possible to interconnect computers in a local area network (LAN), as well as in a wide area network (WAN). The main advantages of DCSs are as follows:

- A higher performance is obtained from a distributed system than from a centralized control system.

- A distributed system is more reliable than a centralized system. In the case of a centralized system, if the computer fails, the whole plant becomes unusable. In a DCS, if one computer fails, only a small part of the plant will be affected and the load of the failed computer can usually be distributed among the other computers.

- A DCS can easily be expanded by adding more computers to the network. For example, if 10 computers are used to control the temperature of 10 ovens, then if the number of ovens is increased to 15, it is easy to add five more computers to the network.

- A DCS is more flexible than a centralized control system as it can be easily adjusted to plant requirements.

In a DCS the sensors and actuators can be connected to local computers which can execute localized controller algorithms. Thus, the local computers in a distributed control environment are usually used for direct digital control (DDC). In a DDC application the computer is used only to carry out the control action for the plant. It is also possible to add some level of supervisory control action to a DDC computer, such as displaying the values of sensors, inputs and outputs.

Distributed control systems are generally used as client–server systems. In such a system one computer (or more if necessary) is designated as the *server* and carries out the common control operations. Other computers in the system are called *clients* and they obey and implement the instructions they receive from the master computer. For example, the task of a client computer could be to receive and format analog data from a sensor, and then pass this data to the server computer every second.

Distributed control systems usually exist within finite boundaries, such as within a factory complex, and all the computers communicate with each other using a LAN cable. Wireless LAN systems are becoming popular, and there is no reason why a DCS cannot be constructed using wireless LAN technology. Using wireless, system reconfiguration is as easy as just adding or removing a computer.

1.4 SCADA SYSTEMS

The term SCADA is an abbreviation for *supervisory control and data acquisition*. SCADA systems integrate the data acquisition and system monitoring and control activities using graphical software packages. A SCADA system is nothing but a customized graphical applications program with all the necessary hardware components. It can be developed using the popular visual programming languages such as *Visual C++* or *Visual Basic*. Good human–computer interface techniques should be employed in the design of the user interface. Alternatively, graphical programming languages such as *Labview* or *VisiDaq* can be used to create powerful, user-friendly SCADA systems.

In a SCADA system the user can have access to a graphical screen in order to monitor or change a setting in the plant. SCADA systems consist of both hardware and software and are usually implemented using personal computers (PCs). Typical hardware includes the computer, keyboard, touch screen, sensors and actuators. The software is in the form of a graphical user interface, where parts of the plant, sensor data and actuator data can all be displayed in various colours on a screen. The advantage of a SCADA system is that the user can easily monitor the status of the overall system. It is important that a SCADA system should be secure and password protected to avoid unauthorized access to the control screens.

1.5 HARDWARE REQUIREMENTS FOR COMPUTER CONTROL

1.5.1 General Purpose Computers

In general, although almost any digital computer can be used for digital control there are some requirements that should be satisfied before a computer is used for such an application. Today, the majority of small and medium scale DDC-type applications are based on microcontrollers which are used as embedded controllers. Applications where user interaction and supervisory control are required are commonly designed around the standard PC hardware.

As shown in Figure 1.8, a general purpose computer consists of the following basic building elements:

- central processing unit (CPU);
- program memory;
- data memory;
- input–output devices.

The CPU is the part which contains the arithmetic and logic unit (ALU), the control unit (CU) and the general purpose registers (GPR). The ALU consists of the logic circuitry necessary to carry out arithmetic and logic operations, for example to add or subtract numbers, to compare numbers and so on. Some ALU units are equipped to carry out multiplication and division and floating point mathematical operations. The CU supervises the operations within the CPU, fetches instructions from the program memory, decodes these instructions and controls the ALU and other parts of the computer so that the required operations can be implemented. The GPR are a set of fast registers which are generally used to carry out fast operations within the CPU.

The program memory of a general purpose computer is usually an external unit and attached to the computer via the data bus and the address bus. A bus is a collection of conductors which carry electrical signals. The data bus is a bidirectional bus which carries the data to be sent or received between the CPU and the other parts of the computer. The size of this bus is 8 bits in most microprocessors and microcontrollers. Some microcontrollers have data buses that are 16 or even 32 bits wide. Minicomputers and mainframe computers usually have 64 or even higher data widths. The address bus is a unidirectional bus which is used to address the peripheral

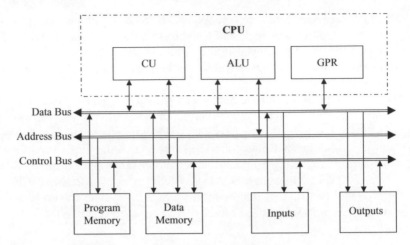

Figure 1.8   Schematic of a general purpose computer

devices attached to the computer. For example, when data is to be written to the memory the address of the memory location is sent on the address bus and the actual data byte is sent on the data bus. The program memory is usually a nonvolatile memory, such as electrically programmable read-only memory (EPROM), EEPROM or flash memory. EPROM memory can be programmed using a suitable programmer device. This type of memory has to be erased using an ultraviolet light source before the contents can be changed. EEPROM memory can be programmed and erased by sending electrical signals to the memory. The disadvantage of this memory is that it is usually a slow process to write or read data from an EEPROM memory. Currently, flash memory is one of the most popular types of nonvolatile memory used. Flash memory is fast and can be erased under program control.

The data memory is usually a volatile memory, used to store the user data. RAM type memories are commonly used for this purpose. The size of this memory can vary from several tens of kilobytes to tens of gigabytes.

Minicomputers and larger computers are equipped with auxiliary storage mediums such as hard disks and magnetic tapes. These devices provide bulk storage for programs and data. Magnetic tape is usually used to store the entire contents of a hard disk for backup purposes.

Input–output devices are also known as the peripheral devices. Many different types of input devices – scanner, camera, keyboard, microphone and mouse – can be connected to the computer. The output devices can be printers, plotters, speakers, visual display units and so on.

General purpose computers are usually more suited to data processing type applications. For example, a minicomputer can be used in an office to provide word processing. Similarly, a large computer can be used in a bank to store and manipulate the accounts of thousands of customers.

1.5.2   Microcontrollers

A microcontroller is a single-chip computer that is specifically manufactured for embedded computer control applications. These devices are very low-cost and can be used very easily in digital control applications. Most microcontrollers have the built-in circuits necessary for computer control applications. For example, a microcontroller may have A/D converters so that the external signals can be sampled. They also have parallel input–output ports so that digital data can be read or output from the microcontroller. Some devices have built-in D/A converters and the output of the converter can be used to drive the plant through an actuator (e.g. an amplifier). Microcontrollers may also have built-in timer and interrupt logic. Using the timer or the interrupt facilities, we can program the microcontroller to implement the control algorithm accurately.

Microcontrollers have traditionally been programmed using the assembly language of the target device. As a result, the assembly languages of the microcontrollers manufactured by different firms are totally different and the user has to learn a new language before being able program a new type of device. Nowadays microcontrollers can be programmed using high-level languages such as BASIC, PASCAL or C. High-level languages offer several advantages compared to the assembly language:

- It is easier to develop programs using a high-level language.
- Program maintenance is much easier if the program is developed using a high-level language.
- Testing a program developed in a high-level language is much easier.
- High-level languages are more user-friendly and less prone to making errors.
- It is easier to document a program developed using a high-level language.

In addition to the above advantages, high-level languages have some disadvantages. For example, the length of the code in memory is usually larger when a high-level language is used, and the programs developed using the assembly language usually run faster than those developed using a high-level language.

In this book, PIC microcontrollers are used as digital controllers. The microcontrollers are programmed using the high-level C language.

1.6   SOFTWARE REQUIREMENTS FOR COMPUTER CONTROL

Computer hardware is nowadays very fast, and control computers are generally programmed using a high-level language. The use of the assembly language is reserved for very special and time-critical applications, such as fast, real-time device drivers. C is a popular language used in most computer control applications. It is a powerful language that enables the programmer to perform low-level operations, without the need to use the assembly language.

The software requirements in a control computer can be summarized as follows:

- the ability to read data from input ports;
- the ability to send data to output ports;
- internal data transfer and mathematical operations;
- timer interrupt facilities for timing the controller algorithm.

All of these requirements can be met by most digital computers, and, as a result, most computers can be used as controllers in digital control systems. The important point is that it is not justified and not cost-effective to use a minicomputer to control the speed of a motor, for example. A microcontroller is much more suitable for this kind of control application. On the other hand, if there are many inputs and many outputs, and if it is required to provide supervisory tasks as well then the use of a minicomputer can easily be justified.

The controller algorithm in a computer is implemented as a program which runs continuously in a loop which is executed at the start of every sampling time. Inside the loop, the desired reference value is read, the actual plant output is also read, and the difference between the desired value and the actual value is calculated. This forms the error signal. The control algorithm is then implemented and the controller output for this sampling instant is calculated. This output is sent to a D/A converter which generates an analog equivalent of the desired control action. This signal is then fed to an actuator which in turn drives the plant to the desired point.

The operation of the controller algorithm, assuming that the reference input and the plant output are digital signals, is summarized below as a sequence of simple steps:

Repeat Forever

When it is time for next sampling instant

- Read the desired value, R
- Read the actual plant output, Y
- Calculate the error signal, $E = R - Y$
- Calculate the controller output, U
- Send the controller output to D/A converter
- Wait for the next sampling instant

End

Similarly, if the reference input and the plant output are analog signals, the operation of the controller algorithm can be summarized as:

Repeat Forever

When it is time for next sampling instant

- Read the desired value, R, from A/D converter
- Read the actual plant output, Y, from the A/D converter
- Calculate the error signal, $E = R - Y$
- Calculate the controller output, U
- Send the controller output to D/A converter
- Wait for the next sampling instant

End

One of the important features of the above algorithms is that once they have been started they run continuously until some event occurs to stop them or until they are stopped manually by an operator. It is important to make sure that the loop is run continuously and exactly at the same times, i.e. exactly at the sampling instants. This is called synchronization and there are several ways in which synchronization can be achieved in practice, such as:

- using polling in the control algorithm;
- using external interrupts for timing;
- using timer interrupts;

- ballast coding in the control algorithm;
- using an external real-time clock.

These methods are discussed briefly here.

1.6.1 Polling

Polling is the software technique where we keep waiting until a certain event occurs, and only then perform the required actions. This way, we wait for the next sampling time to occur and only then run the controller algorithm.

The polling technique is used in DDC applications since the controller cannot do any other operation during the waiting of the next sampling time. The polling technique is described below as a sequence of steps:

Repeat Forever
While Not sampling time
  Wait
End

- Read the desired value, R
- Read the actual plant output, Y
- Calculate the error signal, $E = R - Y$
- Calculate the controller output, U
- Send the controller output to D/A converter

End

1.6.2 Using External Interrupts for Timing

The controller synchronization task can easily be performed using an external interrupt. Here, the controller algorithm can be written as an interrupt service routine (ISR) which is associated with an external interrupt. The external interrupt will typically be a clock with a period equal to the required sampling time. Thus, the computer will run the interrupt service (i.e. the algorithm) routine at every sampling instant. At the end of the ISR control is returned to the main program where the program either waits for the occurrence of the next interrupt or can perform other tasks (e.g. displaying data on a LCD) until the next external interrupt occurs.

The external interrupt approach provides accurate implementation of the control algorithm as far as the sampling time is concerned. One drawback of this method is that an external clock is required to generate the interrupt pulses.

The external interrupt technique has the advantage that the controller is not waiting and can perform other tasks in between the sampling instants.

The external interrupt technique of synchronization is described below as a sequence of steps:

Main program:
  Wait for an external interrupt (or perform some other tasks)
End

Interrupt service routine (ISR):

- Read the desired value, R
- Read the actual plant output, Y
- Calculate the error signal, $E = R - Y$
- Calculate the controller output, U
- Send the controller output to D/A converter

Return from interrupt

1.6.3 Using Timer Interrupts

Another popular way to perform controller synchronization is to use the timer interrupt available on most microcontrollers. Here, the controller algorithm is written inside the timer interrupt service routine, and the timer is programmed to generate interrupts at regular intervals, equal to the sampling time. At the end of the algorithm control returns to the main program, which either waits for the occurrence of the next interrupt or performs other tasks (e.g. displaying data on an LCD) until the next interrupt occurs.

The timer interrupt approach provides accurate control of the sampling time. Another advantage of this technique is that no external hardware is required since the interrupts are generated by the internal timer of the microcontroller.

The timer interrupt technique of synchronization is described below as a sequence of steps:

Main program:
  Wait for a timer interrupt (or perform some other tasks)
End
Interrupt service routine (ISR):

- Read the desired value, R
- Read the actual plant output, Y
- Calculate the error signal, $E = R - Y$
- Calculate the controller output, U
- Send the controller output to D/A converter

Return from interrupt

1.6.4 Ballast Coding

In this technique the loop timing is made to be independent of any external or internal timing signals. The method involves finding the execution time of each instruction inside the loop and then adding *dummy* code to make the loop execution time equal to the required sampling time.

This method has the advantage that no external or internal hardware is required. But one big disadvantage is that if the code inside the loop is changed, or if the CPU clock rate of the microcontroller is changed, then it will be necessary to readjust the execution timing of the loop.

The ballast coding technique of synchronization is described below as a sequence of steps. Here, it is assumed that the loop timing needs to be increased and some dummy code is added to the end of the loop to make the loop timing equal to the sampling time:

Do Forever:

- Read the desired value, R
- Read the actual plant output, Y
- Calculate the error signal, $E = R - Y$
- Calculate the controller output,U
- Send the controller output to D/A converter

  Add dummy code
  . . .
  . . .
  Add dummy code

End

1.6.5 Using an External Real-Time Clock

This technique is similar to using an external interrupt to synchronize the control algorithm. Here, some real-time clock hardware is attached to the microcontroller where the clock is updated at every *tick*; for example, depending on the clock used, 50 ticks will be equal to 1 s if the tick rate is 20 ms. The real-time clock is then read continuously and checked against the time for the next sample. Immediately on exiting from the wait loop the current value of the time is stored and then the time for the next sample is updated by adding the stored time to the sampling interval. Thus, the interval between the successive runs of the loop is independent of the execution time of the loop.

Although the external clock technique gives accurate timing, it has the disadvantage that real-time clock hardware is needed.

The external real-time clock technique of synchronization is described below as a sequence of steps. T is the required sampling time in ticks, which is set to n at the beginning of the algorithm. For example, if the clock rate is 50 Ticks per second, then a Tick is equivalent to 20 ms, and if the required sampling time is 100 ms, we should set $T = 5$:

$T = n$
Next_Sample_Time = Ticks + T
Do Forever:
While Ticks < Next_Sample_Time
  Wait
End
Current_Time = Ticks

- Read the desired value, R
- Read the actual plant output, Y
- Calculate the error signal, $E = R - Y$
- Calculate the controller output, U

- Send the controller output to D/A converter
- Next_Sample_Time = Current_Time + T

End

1.7   SENSORS USED IN COMPUTER CONTROL

Sensors are an important part of closed-loop systems. A sensor is a device that outputs a signal which is related to the measurement of (i.e. is a function of) a physical quantity such as temperature, speed, force, pressure, displacement, acceleration, torque, flow, light or sound. Sensors are used in closed-loop systems in the feedback loops, and they provide information about the actual output of a plant. For example, a speed sensor gives a signal proportional to the speed of a motor and this signal is subtracted from the desired speed reference input in order to obtain the error signal.

Sensors can be classified as analog or digital. Analog sensors are more widely available, and their outputs are analog voltages. For example, the output of an analog temperature sensor may be a voltage proportional to the measured temperature. Analog sensors can only be connected to a computer by using an A/D converter. Digital sensors are not very common and they have logic level outputs which can directly be connected to a computer input port.

The choice of a sensor for a particular application depends on many factors such as the cost, reliability, required accuracy, resolution, range and linearity of the sensor. Some important factors are described below.

Range. The range of a sensor specifies the upper and lower limits of the measured variable for which a measurement can be made. For example, if the range of a temperature sensor is specified as 10–60 °C then the sensor should only be used to measure temperatures within that range.

Resolution. The resolution of a sensor is specified as the largest change in measured value that will not result in a change in the sensor's output, i.e. the measured value can change by the amount quoted by the resolution before this change can be detected by the sensor. In general, the smaller this amount the better the sensor is, and sensors with a wide range have less resolution. For example, a temperature sensor with a resolution of 0.001 K is better than a sensor with a resolution of 0.1 K.

Repeatability. The repeatability of a sensor is the variation of output values that can be expected when the sensor measures the same physical quantity several times. For example, if the voltage across a resistor is measured at the same time several times we may get slightly different results.

Linearity. An ideal sensor is expected to have a linear transfer function, i.e. the sensor output is expected to be exactly proportional to the measured value. However, in practice all sensors exhibit some amount of nonlinearity depending upon the manufacturing tolerances and the measurement conditions.

Dynamic response. The dynamic response of a sensor specifies the limits of the sensor characteristics when the sensor is subject to a sinusoidal frequency change. For example, the dynamic response of a microphone may be expressed in terms of the 3-dB bandwidth of its frequency response.

In the remainder of this chapter, the operation and the characteristics of some of the popular sensors are discussed.

1.7.1 Temperature Sensors

Temperature is one of the fundamental physical variables in most chemical and process control applications. Accurate and reliable measurement of the temperature is important in nearly all process control applications.

Temperature sensors can be analog or digital. Some of the most commonly used analog temperature sensors are: thermocouples, resistance temperature detectors (RTDs) and thermistors. Digital sensors are in the form of integrated circuits. The choice of a sensor depends on the accuracy, the temperature range, speed of response, thermal coupling, the environment (chemical, electrical, or physical) and the cost.

As shown in Table 1.1, thermocouples are best suited to very low and very high temperature measurements. The typical measuring range is from $-270\,°C$ to $+2600\,°C$. In addition, thermocouples are low-cost, very robust, and they can be used in chemical environments. The typical accuracy of a thermocouple is $±1\,°C$. Thermocouples do not require external power for operation.

RTDs are used in medium-range temperature measurements, ranging from $-200\,°C$ to $+600\,°C$. They can be used in most chemical environments but they are not as robust as thermocouples. The typical accuracy of RTDs is $±0.2\,°C$. They require external power for operation.

Thermistors are used in low- to medium-temperature applications, ranging from $-50\,°C$ to about $+200\,°C$. They are not as robust as thermocouples or RTDs and they cannot easily be used in chemical environments. Thermistors are also low-cost devices, they require external power for operation, and they have an accuracy of $±0.2\,°C$.

Integrated circuit temperature sensors are used in low-temperature applications, ranging from $-40\,°C$ to $+125\,°C$. These devices can be either analog or digital, and their coupling with the environment is not very good. The accuracy of integrated circuit sensors is around $±1\,°C$. Integrated temperature sensors differ from other sensors in some important ways:

- They are relatively small.

- Their outputs are highly linear.

- Their temperature range is limited.

- Their cost is very low.

- Some models include advanced features, such as thermostat functions, built-in A/D converters and so on.

- An external power supply is required to operate them.

Table 1.1 Temperature sensors

Sensor	Temperature range ($°C$)	Accuracy ($±°C$)	Cost	Robustness
Thermocouple	-270 to $+2600$	1	Low	Very high
RTD	-200 to $+600$	0.2	Medium	High
Thermistor	-50 to $+200$	0.2	Low	Medium
Integrated circuit	-40 to $+125$	1	Low	Low

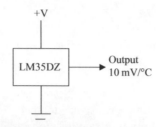

Figure 1.9   LM35DZ temperature sensor

Analog integrated circuit temperature sensors can be voltage output or current output devices. Voltage output sensors give a voltage which is directly proportional to the measured temperature. Similarly, current output sensors act as high-impedance current sources, giving an output current which is proportional to the temperature.

A popular voltage output analog integrated circuit temperature sensor is the LM35DZ, manufactured by National Semiconductors Inc. (see Figure 1.9). This is a 3-pin analog output sensor which provides a linear output voltage of 10 mV/°C. The temperature range is 0 °C to +100 °C, with an accuracy of ±1.5 °C.

The AD590 is an analog integrated circuit sensor with a current output. The device operates in the range −55 °C to +150 °C and produces an output current of 1 μA/°C.

Digital integrated temperature sensors produce digital outputs which can be interfaced to a computer. The output data format is usually nonstandard and the measured temperature can be extracted by using suitable algorithms. The DS1620 is a popular digital temperature sensor which also incorporates digitally programmable thermostat outputs. The device provides a 9-bit serial data to indicate the measured temperature. Data is extracted from the device by sending clock pulses and then reading the data after each pulse. Table 1.2 shows the sensor's measured temperature–output relationship.

There can be several sources of error during the measurement of temperature. Some important possible errors are described below.

Sensor self-heating. RTDs, thermistors and integrated circuit sensors require an external power supply for their operation. The power supply can cause the sensor to heat, leading to an error in the measurement. The effect of self-heating depends on the size of the sensor and the amount of power dissipated by the sensor. Self-heating can be avoided by using the lowest possible external power, or by considering the heating effect in the measurement.

Table 1.2   Temperature–data relationship of DS1620

Temperature (°C)	Digital output
+125	0 11111010
+25	0 00110010
0.5	0 00000001
0	0 00000000
−0.5	1 11111111
−25	1 11001110
−55	1 10010010

Electrical noise. Electrical noise can introduce errors into the measurement. Thermocouples produce very low voltages (of the order of tens of microvolts) and, as a result, noise can easily enter the measurement. This noise can usually be minimized by using low-pass filters, and by keeping the sensor leads as short as possible and away from motors and other electrical machinery.

Mechanical stress. Some sensors such as RTDs are sensitive to mechanical stress and should be used carefully. Mechanical stress can be minimized by avoiding deformation of the sensor.

Thermal coupling. It is important that for accurate and fast measurements the sensor should make a good contact with the measuring surface. If the surface has a thermal gradient then incorrect placement of the sensor can lead to errors. If the sensor is used in a liquid, the liquid should be stirred to cause a uniform heat distribution. Integrated circuit sensors usually suffer from thermal coupling since they are not easily mountable on surfaces.

Sensor time constant. The response time of the sensor can be another source of error. Every type of sensor takes a finite time to respond to a change in its environment. Errors due to the sensor time constant can be minimized by improving the coupling between the sensor and the measuring surface.

1.7.2 Position sensors

Position sensors are used to measure the position of moving objects. These sensors are basically of two types: sensors to measure linear movement, and sensors to measure angular movement.

Potentiometers are available in linear and rotary forms. In a typical application, a fixed voltage is applied across the potentiometer and the voltage across the potentiometer arm is measured. This voltage is proportional to the position of the arm, and hence by measuring the voltage we know the position of the arm. Figure 1.10 shows a linear potentiometer. If the applied voltage is V_i, the voltage across the arm is given by

$$V_a = kV_i y$$

where y is the position of the arm from the beginning of the potentiometer, and k is a constant.

Figure 1.11 shows a rotary potentiometer which can be used to measure angular position. If V_i is again the applied voltage, the voltage across the arm is given by

$$V_a = kV_i \theta$$

where θ is the angle of the arm, and k is a constant.

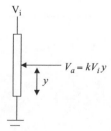

Figure 1.10   Linear potentiometer

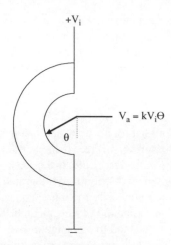

Figure 1.11   Rotary potentiometer

Potentiometer type position sensors are low-cost, but they have the disadvantage that the range is limited and also that the sensor can be worn out by excessive movement of the arm.

Among other types of position sensors are capacitive sensors, inductive sensors, linear variable differential transformers (LVDTs) and optical encoders. Capacitive position sensors rely on the fact that the capacitance of a parallel plate capacitor changes as the distance between the plates is changed. The formula for the capitance, C, of a parallel plate capacitor is

$$C = \varepsilon \frac{A}{d},$$ (1.1)

where ε is the dielectric constant, A the area of the plates and d the distance between the plates.

Typically, the capacitor of the sensor is used in the feedback loop of an operational amplifier as shown in Figure 1.12, and a reference capacitor is used at the input. If a voltage V_i is applied, the output voltage V_o is given by

$$V_o = -V_i \frac{C}{C_{\text{ref}}}.$$ (1.2)

Using equation (1.1), we obtain

$$V_o = -V_i \frac{C_{\text{ref}} d}{\varepsilon A}.$$ (1.3)

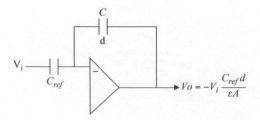

Figure 1.12   Position sensor using capacitors

Figure 1.13   Commercially available LVDT sensor

From (1.3), if we apply a constant amplitude sinusoidal signal as the input, the amplitude of the output voltage is proportional to the distance between the plates.

LVDT sensors (see Figure 1.13) consist of one primary and two secondary windings on a hollow cylinder. The primary winding is in the middle, and the secondary windings have equal number of turns, series coupled, and they are at the ends of the cylinder (see Figure 1.14). A sinusoidal signal with a voltage of 0.5–5 V and frequency 1–20 kHz is applied to the primary winding. A magnetic core which measures the position moves inside the cylinder, and the movement of this core varies the magnetic field linking the primary winding to the secondary windings. Because the secondary windings are in opposition, the movement of the core to one position increases the induced voltage in one secondary coil and decreases the induced voltage in the other secondary coil. The net voltage difference is proportional to the position of the core inside the cylinder. Thus, by measuring the induced voltage we know the position of the core. The strong relationship between the core position and the induced voltage yields a design that exhibits excellent resolution. Most commercially available LVDTs come with built-in signal-conditioning circuitry that provides an easy interface to a computer. The device operates from a d.c. supply and the signal conditioner provides the a.c. signal required for the operation of the circuit, as well as the demodulation of the output signal to give a useful d.c.

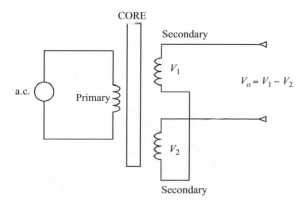

Figure 1.14   LVDT sensor circuit diagram

voltage output. The range of an LVDT is from $\pm 125 \, \mu$m to ± 75 mm and the sensitivity ranges from 0.6 to 30 mV per 25 μm under normal excitation of 3–6 V.

The advantages of LVDT are:

- low cost;
- robust design;
- no hysteresis effect;
- fast response time;
- no friction resistance;
- long life.

The main disadvantage of the LVDT is that the core must have direct contact with the measured surface, which may not always be possible.

1.7.3 Velocity and acceleration sensors

Velocity is the differentiation of position, and in general position sensors can be used to measure velocity. The required differentiation can be done either in hardware (e.g. using operational amplifiers) or by the computer. For more accurate measurements velocity sensors should be used. There are two types of velocity sensors: linear sensors, and rotary sensors.

Linear velocity sensors can be constructed using a pair of coils and a moving magnet. When the coils are connected in series, the movement of the magnet produces additive voltage which is proportional to the movement of the magnet.

One of the most widely used rotary velocity sensors is the tachometer (or tachogenerator). A tachometer (see Figure 1.15) is connected to the shaft of a rotating device (e.g. a motor) and produces an analog d.c. voltage which is proportional to the speed of the shaft. If ω is the angular velocity of the shaft, the output voltage of the tachometer is given by

$$V_o = k\omega,$$

where k is the gain constant of the tachometer.

Another popular velocity sensor is the optical encoder. This basically consists of a light source and a disk with opaque and transparent sections where the disk is attached to the rotating shaft. A light sensor at the other side of the wheel detects light and a pulse is produced when the transparent section of the disk comes round. The encoder's controller counts the pulses in a given time, and this is proportional to the speed of the shaft. Figure 1.16 shows a typical commercial encoder.

Figure 1.15 Commercially available tachometer

Figure 1.16 Commercially available encoder

Acceleration is the differentiation of velocity, or the double differentiation of position. Thus, in general, position sensors can be used to measure acceleration. The differentiation can be done either by using operational amplifiers or by a computer program. For accurate measurement of the acceleration, semiconductor accelerometers can be used. For example, the ADXL202 is an accelerometer chip manufactured by Analog Devices Inc. This is a low-cost 8-pin chip with two outputs to measure the acceleration in two dimensions. The outputs are digital signals whose duty cycles are proportional to the acceleration in each of the two axes. These outputs can be connected directly to a microcontroller and the acceleration can be measured very easily, requiring no A/D converter. The measurement range of the ADXL202 is $\pm 2\,g$, where g is acceleration due to gravity, and the device can measure both dynamic acceleration (e.g. vibration), and static acceleration (e.g. gravity).

1.7.4 Force sensors

Force sensors can be constructed using position sensors. Alternatively, a strain gauge can be used to measure force accurately. There are many different types of strain gauges. A strain gauge can be made from capacitors and inductors, but the most widely used types are made from resistors. A wire strain gauge is made from a resistor, in the form of a metal foil. The principle of operation is that the resistance of a wire increases with increasing strain and decreases with decreasing strain.

In order to measure strain with a strain gauge, it must be connected to an electrical circuit, and a Wheatstone bridge is commonly used to detect the small changes in the resistance of the strain gauge.

Strain gauges can be used to measure force, load, weight pressure, torque or displacement.

Force can also be measured using the principle of piezoelectricity. A piezoelectric sensor produces voltage when a force is applied to its surface. The disadvantage of this method is that the voltage decays after the application of the force and thus piezoelectric sensors are only useful for measuring dynamic force.

1.7.5 Pressure sensors

Early pressure measurement was based on using a flexible device (e.g. a diaphragm) as a sensor; the pressure changed as the device moved and caused a dial connected to the device to move and indicate the pressure. Nowadays, the movement is converted into an electrical signal which is proportional to the applied pressure. Strain gauges, capacitance change, inductance change, piezoelectric effect, optical pressure sensors and similar techniques are used to measure the pressure.

1.7.6 Liquid sensors

There are many different types of liquid sensors. These sensors are used to:

- detect the presence of liquid;
- measure the level of liquid;
- measure the flow rate of liquid, for example through a pipe.

The presence of a liquid can be detected by using optical, ultrasonic, change of resistance, change of capacitance or similar techniques. For example, optical technique is based on using an LED and a photo-transistor, both housed within a plastic dome and at the head of the device. When no liquid is present, light from the LED is internally reflected from the dome to the photo-transistor and the output is designed to be off. When liquid is present the dome is covered with liquid and the refractive index at the dome–liquid boundary changes, allowing some light to escape from the LED. As a result of this, the amount of light received by the photo-transistor is reduced and the output is designed to switch on, indicating the presence of liquid.

The level of liquid in a tank can be measured using immersed sensor techniques, or nontouching ultrasonic techniques. The simplest technique is to immerse a rod in the liquid with a potentiometer placed inside the rod. The potentiometer arm is designed to move as the level of the liquid is changed. The change in the resistance can be measured and hence the level of the liquid is obtained.

Pressure sensors are also used to measure the level of liquid in a tank. Typically, the pressure sensor is mounted at the bottom of the tank where change of pressure is proportional to the height of the liquid. These sensors usually give an analog output voltage proportional to the height of the liquid inside the tank.

Nontouching ultrasonic level measurement is very accurate, but more expensive than the other techniques. Basically, an ultrasonic beam is sent to the surface of the water and the echo of the beam is detected. The time difference between sending the beam and the echo is proportional to the level of the liquid in the tank.

The liquid flow rate can be measured by several techniques:

- paddlewheel sensors;
- displacement flow meters;
- magnetic flow meters;

Paddlewheel sensors are cost-effective and very popular for the measurement of liquid flow rate. A wheel is mounted inside the sensor whose speed of rotation is proportional to the flow rate. As the wheel rotates a voltage is produced which indicates the flow rate.

Displacement flow meters measure the flow rate of a liquid by separating the flow into known volumes and counting them over time. These meters provide good accuracy. Displacement flow meters have several types such as sliding vane meters, rotary piston meters, helix flow meters and so on.

Figure 1.17   Commercially available magnetic flow rate sensor (Sparling Instruments Inc.)

Magnetic flow meters are based on Faraday's law of magnetic induction. Here, the liquid acts as a conductor as it flows through a pipe. This induces a voltage which is proportional to the flow rate. The faster the flow rate, the higher is the voltage. This voltage is picked up by the sensors mounted in the meter tube and electronic means are used to calculate the flow rate based on the cross-sectional area of the tube. Advantages of magnetic flow rates are as follows:

- Corrosive liquids can be used.
- The measurement does not change the flow stream.

Figure 1.17 shows a typical magnetic flow meter.

1.7.7   Air flow sensors

Air flow is usually measured using anemometers. A classical anemometer (see Figure 1.18) has a rotating vane, and the speed of rotation is proportional to the air flow. Hot wire anemometers

Figure 1.18   Classical anemometer

Figure 1.19   Hot wire anemometer (Extech Instruments Corp.)

have no moving parts (Figure 1.19). The sensor consists of an electrically heated platinum wire which is placed in the air flow. As the flow velocity increases the rate of heat flow from the heated wire to the flow stream increases and a cooling occurs on the electrode, causing its resistance to change. The flow rate is then determined from the change in the resistance.

1.8 EXERCISES

1. Describe what is meant by accuracy, range and resolution.

2. Give an example of how linear displacement can be measured.

3. A tachometer is connected to a motor shaft in a speed control system. If the tachometer produces 100 mV per revolution, write an expression for its transfer function in terms of volts/radians-per-second.

4. What is polling in software? Describe how a control algorithm can be synchronized using polling.

5. What are the differences between polling and interrupt based processing? Which method is more suitable in digital controller design.

6. Explain why an A/D converter may be required in digital control systems.

7. Explain the operation of an LVDT sensor. You are required to design a motor position control system. Explain what type of position transducer you would use in your design.

8. You are required to measure the flow rate of water entering into a tank. Explain what type of sensor you can use.

9. Water enters into a tank through a pipe. At the same time, a certain amount of water is output from the tank continuously. You are required to design a water level control system so that the level of water in the tank is kept constant at all times. Draw a sketch of a suitable control system. Explain what types of sensors you will be using in the design.

10. What is a paddlewheel? Explain the operation principles of a paddlewheel liquid flow meter. What are the advantages and disadvantages of this sensor?

11. Compare the vane based anemometer and the hot air anemometer. Which one would you choose to measure the air flow through a narrow pipe?

12. Explain the factors that should be considered before purchasing and using a sensor.

FURTHER READING

[Bennett, 1994]   Bennett, S. Real-time Computer Control: An Introduction. Prentice Hall, Hemel Hempstead, 1994.

[D'Souza, 1988] D'Souza, A.F. Design of Control Systems. Prentice Hall, Englewood Cliffs, NJ, 1988.

[Nise, 2000]   Nise, N.S. Control Systems Engineering, 3rd edn., John Wiley & Sons, Inc., New York, 2000.

2

System Modelling

The task of mathematical modelling is an important step in the analysis and design of control systems. In this chapter, we will develop mathematical models for the mechanical, electrical, hydraulic and thermal systems which are used commonly in everyday life. The mathematical models of systems are obtained by applying the fundamental physical laws governing the nature of the components making these systems. For example, Newton's laws are used in the mathematical modelling of mechanical systems. Similarly, Kirchhoff's laws are used in the modelling and analysis of electrical systems.

Our mathematical treatment will be limited to linear, time-invariant ordinary differential equations whose coefficients do not change in time. In real life many systems are nonlinear, but they can be linearized around certain operating ranges about their equilibrium conditions. Real systems are usually quite complex and exact analysis is often impossible. We shall make approximations and reduce the system components to idealized versions whose behaviours are similar to the real components.

In this chapter we shall look only at the passive components. These components are of two types: those storing energy (e.g. the capacitor in an electrical system), and those dissipating energy (e.g. the resistor in an electrical system).

The mathematical model of a system is one or more differential equations describing the dynamic behaviour of the system. The Laplace transformation is applied to the mathematical model and then the model is converted into an algebraic equation. The properties and behaviour of the system can then be represented as a block diagram, with the transfer function of each component describing the relationship between its input and output behaviour.

2.1 MECHANICAL SYSTEMS

Models of mechanical systems are important in control engineering because a mechanical system may be a vehicle, a robot arm, a missile, or any other system which incorporates a mechanical component. Mechanical systems can be divided into two categories: translational systems and rotational systems. Some systems may be purely translational or rotational, whereas others may be hybrid, incorporating both translational and rotational components.

Microcontroller Based Applied Digital Control   D. Ibrahim
© 2006 John Wiley & Sons, Ltd

2.1.1 Translational Mechanical Systems

The basic building blocks of translational mechanical systems are masses, springs, and dashpots (Figure 2.1). The input to a translational mechanical system may be a force, F, and the output the displacement, y.

Springs store energy and are used in most mechanical systems. As shown in Figure 2.2, some springs are hard, some are soft, and some are linear. A hard or a soft spring can be linearized for small deviations from its equilibrium condition. In the analysis in this section, a spring is assumed massless, or of negligible mass, i.e. the forces at both ends of the spring are assumed to be equal in magnitude but opposite in direction.

For a linear spring, the extension y is proportional to the applied force F and we have

$$F = ky, \qquad (2.1)$$

where k is known as the *stiffness* constant. The spring when stretched stores energy given by

$$E = \frac{1}{2}ky^2. \qquad (2.2)$$

This energy is released when the spring contracts back to its original length.

In some applications springs can be in parallel or in series. When n springs are in parallel, then the equivalent stiffness constant k_{eq} is equal to the sum of all the individual spring stiffnesses k_i:

$$k_{eq} = k_1 + k_2 + \cdots + k_n. \qquad (2.3)$$

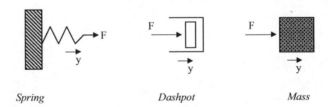

| *Spring* | *Dashpot* | *Mass* |

Figure 2.1   Translational mechanical system components

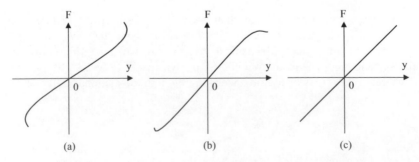

(a)   (b)   (c)

Figure 2.2   (a) Hard spring, (b) soft spring, (c) linear spring

Similarly, when n springs are in series, then the reciprocal of the equivalent stiffness constant k_{eq} is equal to the sum of all the reciprocals of the individual spring stiffnesses k_i:

$$\frac{1}{k_{eq}} = \frac{1}{k_1} + \frac{1}{k_2} + \ldots + \frac{1}{k_n}. \tag{2.4}$$

As an example, if there are two springs k_1 and k_2 in series, then the equivalent stiffness constant is given by

$$k_{eq} = \frac{k_1 k_2}{k_1 + k_2}. \tag{2.5}$$

A *dashpot* element is a form of damping and can be considered to be represented by a piston moving in a viscous medium in a cylinder. As the piston moves the liquid passes through the edges of the piston, damping to the movement of the piston. The force F which moves the piston is proportional to the velocity of the piston movement. Thus,

$$F = c\frac{dy}{dt}. \tag{2.6}$$

A dashpot does not store energy.

When a force is applied to a *mass*, the relationship between the force F and the acceleration a of the mass is given by Newton's second law as $F = ma$. Since acceleration is the rate of change of velocity and the velocity is the rate of change of displacement, we can write

$$F = m\frac{d^2 y}{dt^2}. \tag{2.7}$$

The energy stored in a mass when it is moving is the kinetic energy which is dependent on the velocity of the mass and is given by

$$E = \frac{1}{2}mv^2. \tag{2.8}$$

This energy is released when the mass stops.

Some examples of translational mechanical system models are given below.

Example 2.1

Figure 2.3 shows a simple mechanical translational system with a mass, spring and dashpot. A force F is applied to the system. Derive a mathematical model for the system.

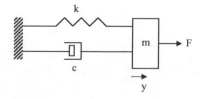

Figure 2.3   Mechanical system with mass, spring and dashpot

Solution

As shown in Figure 2.3, the net force on the mass is the applied force minus the forces exerted by the spring and the dashpot. Applying Newton's second law, we can write

$$F - ky - c\frac{dy}{dt} = m\frac{d^2y}{dt^2} \tag{2.9}$$

or

$$F = m\frac{d^2y}{dt^2} + c\frac{dy}{dt} + ky. \tag{2.10}$$

Equation (2.10) is usually written in the form

$$F = m\ddot{y} + c\dot{y} + ky. \tag{2.11}$$

Taking the Laplace transform of (2.11), we can derive the transfer function of the system as

$$F(s) = ms^2Y(s) + csY(s) + kY(s)$$

or

$$\frac{Y(s)}{F(s)} = \frac{1}{ms^2 + cs + k}. \tag{2.12}$$

The transfer function in (2.12) is represented by the block diagram shown in Figure 2.4.

Example 2.2

Figure 2.5 shows a mechanical system with two masses and two springs. Drive an expression for the mathematical model of the system.

Solution

Applying Newton's second law to the mass m_1,

$$-k_2(y_1 - y_2) - c\left(\frac{dy_1}{dt} - \frac{dy_2}{dt}\right) - k_1y_1 = m_1\frac{d^2y_1}{dt^2}, \tag{2.13}$$

and for the mass m_2,

$$F - k_2(y_2 - y_1) - c\left(\frac{dy_2}{dt} - \frac{dy_1}{dt}\right) = m_2\frac{d^2y_2}{dt^2}, \tag{2.14}$$

we can write (2.13) and (2.14) as

$$m_1\ddot{y}_1 + c\dot{y}_1 - c\dot{y}_2 + (k_1 + k_2)y_1 - k_2y_2 = 0, \tag{2.15}$$

$$m_2\ddot{y}_2 + c\dot{y}_2 - c\dot{y}_1 + k_2y_2 - k_2y_1 = F. \tag{2.16}$$

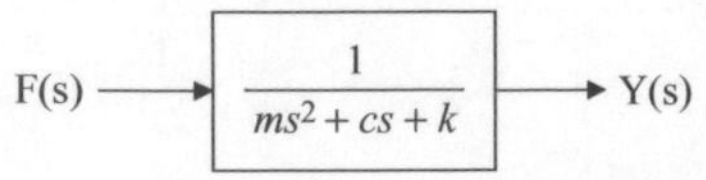

$$F(s) \longrightarrow \boxed{\frac{1}{ms^2 + cs + k}} \longrightarrow Y(s)$$

Figure 2.4 Block diagram of the simple mechanical system

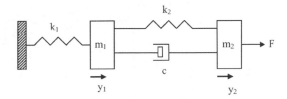

Figure 2.5 Example mechanical system

Equations (2.15) and (2.16) can be written in matrix form as

$$\begin{bmatrix} m_1 & 0 \\ 0 & m_2 \end{bmatrix}\begin{bmatrix} \ddot{y}_1 \\ \ddot{y}_2 \end{bmatrix} + \begin{bmatrix} c & -c \\ -c & c \end{bmatrix}\begin{bmatrix} \dot{y}_1 \\ \dot{y}_2 \end{bmatrix} + \begin{bmatrix} k_1+k_2 & -k_2 \\ -k_2 & k_2 \end{bmatrix}\begin{bmatrix} y_1 \\ y_2 \end{bmatrix} = \begin{bmatrix} 0 \\ F \end{bmatrix}. \tag{2.17}$$

Example 2.3

Figure 2.6 shows a mechanical system with two masses, and forces applied to each mass. Drive an expression for the mathematical model of the system.

Solution

Applying Newton's second law to the mass m_1,

$$F_1 - k(y_1 - y_2) - c\left(\frac{dy_1}{dt} - \frac{dy_2}{dt}\right) = m_1\frac{d^2y_1}{dt^2}, \tag{2.18}$$

and to the mass m_2,

$$F_2 - k(y_2 - y_1) - c\left(\frac{dy_2}{dt} - \frac{dy_1}{dt}\right) = m_2\frac{d^2y_2}{dt^2}, \tag{2.19}$$

we can write (2.13) and (2.14) as

$$m_1\ddot{y}_1 + c\dot{y}_1 - c\dot{y}_2 + ky_1 - ky_2 = F_1, \tag{2.20}$$

$$m_2\ddot{y}_2 + c\dot{y}_2 - c\dot{y}_1 + ky_2 - ky_1 = F_2. \tag{2.21}$$

Equations (2.20) and (2.21) can be written in matrix form as

$$\begin{bmatrix} m_1 & 0 \\ 0 & m_2 \end{bmatrix}\begin{bmatrix} \ddot{y}_1 \\ \ddot{y}_2 \end{bmatrix} + \begin{bmatrix} c & -c \\ -c & c \end{bmatrix}\begin{bmatrix} \dot{y}_1 \\ \dot{y}_2 \end{bmatrix} + \begin{bmatrix} k & -k \\ -k & k \end{bmatrix}\begin{bmatrix} y_1 \\ y_2 \end{bmatrix} = \begin{bmatrix} F_1 \\ F_2 \end{bmatrix}. \tag{2.22}$$

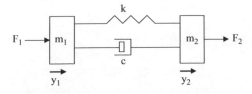

Figure 2.6 Example mechanical system

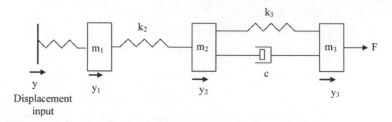

Figure 2.7   Example mechanical system

Example 2.4

Figure 2.7 shows a mechanical system with three masses, two springs and a dashpot. A force is applied to mass m_3 and a displacement is applied to spring k_1. Drive an expression for the mathematical model of the system.

Solution

Applying Newton's second law to the mass m_1,

$$k_1 y - k_1 y_1 - k_2(y_1 - y_2) + k_2 y_2 = m_1 \frac{d^2 y_1}{dt^2} \tag{2.23}$$

to the mass m_2,

$$-c\left(\frac{dy_2}{dt} - \frac{dy_3}{dt}\right) - k_2(y_2 - y_1) - k_3(y_2 - y_3) = m_2 \frac{d^2 y_2}{dt^2}, \tag{2.24}$$

and to the mass m_3,

$$F - c\left(\frac{dy_3}{dt} - \frac{dy_2}{dt}\right) - k_3(y_3 - y_2) = m_3 \frac{d^2 y_3}{dt^2}, \tag{2.25}$$

we can write (2.23)–(2.25) as

$$m_1 \ddot{y}_1 + (k_1 + k_2)y_1 - k_2 y_2 = k_1 y, \tag{2.26}$$

$$m_2 \ddot{y}_2 + c\dot{y}_2 - c\dot{y}_3 - k_2 y_1 + (k_2 + k_3)y_2 - k_3 y_3 = 0, \tag{2.27}$$

$$m_3 \ddot{y}_3 + c\dot{y}_3 - c\dot{y}_2 + k_3 y_3 - k_3 y_2 = F. \tag{2.28}$$

The above equations can be written in matrix form as

$$\begin{bmatrix} m_1 & 0 & 0 \\ 0 & m_2 & 0 \\ 0 & 0 & m_3 \end{bmatrix} \begin{bmatrix} \ddot{y}_1 \\ \ddot{y}_2 \\ \ddot{y}_3 \end{bmatrix} + \begin{bmatrix} 0 & 0 & 0 \\ 0 & c & -c \\ 0 & -c & c \end{bmatrix} \begin{bmatrix} \dot{y}_1 \\ \dot{y}_2 \\ \dot{y}_3 \end{bmatrix} + \begin{bmatrix} k_1 + k_2 & -k_2 & 0 \\ -k_2 & k_2 + k_3 & -k_3 \\ 0 & -k_3 & k_3 \end{bmatrix} \begin{bmatrix} y_1 \\ y_2 \\ y_3 \end{bmatrix} = \begin{bmatrix} k_1 y \\ 0 \\ F \end{bmatrix}.$$

$$\tag{2.29}$$

2.1.2   Rotational Mechanical Systems

The basic building blocks of rotational mechanical systems are the moment of inertia, torsion spring (or rotational spring) and rotary damper (Figure 2.8). The input to a rotational mechanical system may be the torque, T, and the output the rotational displacement, or angle, θ.

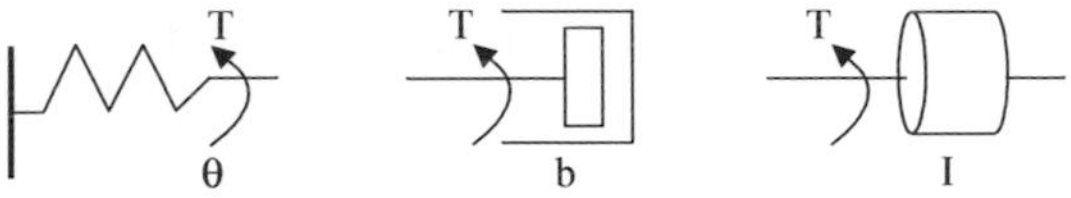

Torsional spring   Rotational dashpot   Moment of inertia

Figure 2.8   Rotational mechanical system components

A *rotational spring* is similar to a translational spring, but here the spring is twisted. The relationship between the applied torque, T, and the angle θ rotated by the spring is given by

$$T = k\theta, \tag{2.30}$$

where θ is known as the rotational *stiffness* constant. In our modelling we are assuming that the mass of the spring is negligible and the spring is linear.

The energy stored in a torsional spring when twisted by an angle θ is given by

$$E = \frac{1}{2}k\theta^2. \tag{2.31}$$

A *rotary damper* element creates damping as it rotates. For example, when a disk rotates in a fluid we get a rotary damping effect. The relationship between the applied torque, T, and the angular velocity of the rotary damper is given by

$$T = c\omega = c\frac{d\theta}{dt}. \tag{2.32}$$

In our modelling the mass of the rotary damper will be neglected, or will be assumed to be negligible. A rotary damper does not store energy.

Moment of inertia refers to a rotating body with a mass. When a torque is applied to a body with a moment of inertia we get an angular acceleration, and this acceleration rotates the body. The relationship between the applied torque, T, angular acceleration, a, and the moment of inertia, I, I is given by

$$T = Ia = I\frac{d\omega}{dt} \tag{2.33}$$

or, since $\omega = d\theta/dt$,

$$T = I\frac{d^2\theta}{dt^2}. \tag{2.34}$$

The energy stored in a mass rotating with an angular velocity ω is given by

$$E = \frac{1}{2}I\omega^2. \tag{2.35}$$

Some examples of rotational system models are given below.

Example 2.5

A disk of moment of inertia I is rotated (see Figure 2.9) with an applied torque of T. The disk is fixed at one end through an elastic shaft. Assuming that the shaft can be modelled with a

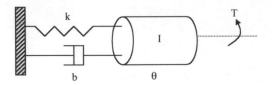

Figure 2.9   Rotational mechanical system

rotational dashpot and a rotational spring, derive an equation for the mathematical model of this system.

Solution

The damper torque and spring torque oppose the applied torque. If θ is the angular displacement from the equilibrium, we can write

$$T - b\frac{d\theta}{dt} - k\theta = I\frac{d^2\theta}{dt^2} \tag{2.36}$$

or

$$I\frac{d^2\theta}{dt^2} + b\frac{d\theta}{dt} + k\theta = T. \tag{2.37}$$

Equation (2.37) is normally written in the form

$$I\ddot{\theta} + b\dot{\theta} + k\theta = T. \tag{2.38}$$

Example 2.6

Figure 2.10 shows a rotational mechanical system with two moments of inertia and a torque applied to each one. Derive a mathematical model for the system.

Solution

For the system shown in Figure 2.10 we can write the following equations: for disk 1,

$$T_1 - k(\theta_1 - \theta_2) - b\left(\frac{d\theta_1}{dt} - \frac{d\theta_2}{dt}\right) = I_1\frac{d^2\theta_1}{dt^2}; \tag{2.39}$$

and for disk 2,

$$T_2 - k(\theta_2 - \theta_1) - b\left(\frac{d\theta_2}{dt} - \frac{d\theta_1}{dt}\right) = I_2\frac{d^2\theta_2}{dt^2}. \tag{2.40}$$

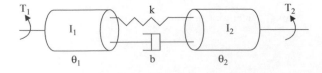

Figure 2.10   Rotational mechanical system

Equations (2.39) and (2.40) can be written as

$$I_1\ddot{\theta}_1 + b\dot{\theta}_1 - b\dot{\theta}_2 + k\theta_1 - k\theta_2 = T_1 \tag{2.41}$$

and

$$I_2\ddot{\theta}_2 - b\dot{\theta}_1 + b\dot{\theta}_2 - k\theta_1 + k\theta_2 = T_2. \tag{2.42}$$

Writing the equations in matrix form, we have

$$\begin{bmatrix} I_1 & 0 \\ 0 & I_2 \end{bmatrix} \begin{bmatrix} \ddot{\theta}_1 \\ \ddot{\theta}_2 \end{bmatrix} + \begin{bmatrix} b & -b \\ -b & b \end{bmatrix} \begin{bmatrix} \dot{\theta}_1 \\ \dot{\theta}_2 \end{bmatrix} + \begin{bmatrix} k & -k \\ -k & k \end{bmatrix} \begin{bmatrix} \theta_1 \\ \theta_2 \end{bmatrix} = \begin{bmatrix} T_1 \\ T_2 \end{bmatrix}. \tag{2.43}$$

2.1.2.1 Rotational Mechanical Systems with Gear-Train

Gear-train systems are very important in many mechanical engineering systems. Figure 2.11 shows a simple gear-train, consisting of two gears, each connected to two masses with moments of inertia I_1 and I_2. Suppose that gear 1 has n_1 teeth and radius r_1, and that gear 2 has n_2 teeth and radius r_2. In this analysis we assume that the gears have no backlash, they are rigid bodies, and the moment of inertia of the gears is assumed to be negligible.

The rotational displacement of the two gears depends on their radii and is given by the relationship

$$r_1\theta_1 = r_2\theta_2 \tag{2.44}$$

or

$$\theta_2 = \frac{r_1}{r_2}\theta_1, \tag{2.45}$$

where θ_1 and θ_2 are the rotational displacements of gear 1 and gear 2, respectively.

The ratio of the teeth numbers is equal to the ratio of the radii and is given by

$$\frac{r_1}{r_2} = \frac{n_1}{n_2} = n, \tag{2.46}$$

where n is the gear teeth ratio.

Assuming that a torque T is applied to the system, we can write

$$I_1\frac{d^2\theta_1}{dt^2} = T - T_1 \tag{2.47}$$

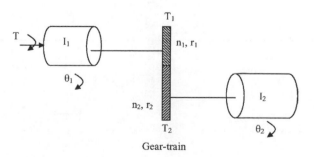

Figure 2.11 A two gear-train system

and

$$I_2 \frac{d^2\theta_2}{dt^2} = T_2. \tag{2.48}$$

Equating the power transmitted by the gear-train,

$$T_1 \frac{d\theta_1}{dt} = T_2 \frac{d\theta_2}{dt} \quad \text{or} \quad \frac{T_1}{T_2} = \frac{d\theta_2/dt}{d\theta_1/dt} = n. \tag{2.49}$$

Substituting (2.49) into (2.47), we obtain

$$I_1 \frac{d^2\theta_1}{dt^2} = T - nT_2 \tag{2.50}$$

or

$$I_1 \frac{d^2\theta_1}{dt^2} = T - n\left(I_2 \frac{d^2\theta_2}{dt^2}\right); \tag{2.51}$$

then, since $\theta_2 = n\theta_1$, we obtain

$$(I_1 + n^2 I_2)\frac{d^2\theta_1}{dt^2} = T. \tag{2.52}$$

It is clear from (2.52) that the moment of inertia of the load, I_2, is reflected to the other side of the gear-train as $n^2 I_2$.

An example of a system coupled with a gear-train is given below.

Example 2.7

Figure 2.12 shows a rotational mechanical system coupled with a gear-train. Derive an expression for the model of the system.

Solution

Assuming that a torque T is applied to the system, we can write

$$I_1 \frac{d^2\theta_1}{dt^2} + b_1 \frac{d\theta_1}{dt} + k_1\theta_1 = T - T_1 \tag{2.53}$$

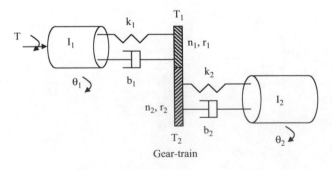

Figure 2.12   Mechanical system with gear-train

and

$$I_2 \frac{d^2\theta_2}{dt^2} + b_2 \frac{d\theta_2}{dt} + k_2\theta_2 = T_2. \tag{2.54}$$

Equating the power transmitted by the gear-train,

$$T_1 \frac{d\theta_1}{dt} = T_2 \frac{d\theta_2}{dt} \quad \text{or} \quad \frac{T_1}{T_2} = \frac{d\theta_2/dt}{d\theta_1/dt} = n. \tag{2.55}$$

Substituting (2.55) into (2.53), we obtain

$$I_1 \frac{d^2\theta_1}{dt^2} + b_1 \frac{d\theta_1}{dt} + k_1\theta_1 = T - nT_2 \tag{2.56}$$

or

$$I_1 \frac{d^2\theta_1}{dt^2} + b_1 \frac{d\theta_1}{dt} + k_1\theta_1 = T - n\left(I_2 \frac{d^2\theta_2}{dt^2} + b_2 \frac{d\theta_2}{dt} + k_2\theta_2 \right). \tag{2.57}$$

Since $\theta_2 = n\theta_1$, this gives

$$(I_1 + n^2 I_2) \frac{d^2\theta_1}{dt^2} + (b_1 + n^2 b_2) \frac{d\theta_1}{dt} + (k_1 + n^2 k_2)\theta_1 = T. \tag{2.58}$$

2.2 ELECTRICAL SYSTEMS

The basic building blocks of electrical systems are the resistor, inductor and capacitor (Figure 2.13). The input to an electrical system may be the voltage, V, and current, i.

The relationship between the voltage across a *resistor* and the current through it is given by

$$V_r = Ri, \tag{2.59}$$

where R is the resistance.

For an *inductor*, the potential difference across the inductor depends on the rate of change of current through the inductor, given by

$$v_L = L \frac{di}{dt}, \tag{2.60}$$

where L is the inductance. Equation (2.60) can also be written as

$$i = \frac{1}{L} \int v_L dt. \tag{2.61}$$

The energy stored in an inductor is given by

$$E = \frac{1}{2} L i^2. \tag{2.62}$$

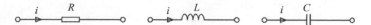

Figure 2.13   Electrical system components

The potential difference across a *capacitor* depends on the charge the plates hold, and is given by

$$v_C = \frac{q}{C}. \tag{2.63}$$

The relationship between the current through the capacitor and the voltage across it is given by

$$i = C \frac{dv_C}{dt} \tag{2.64}$$

or

$$v_C = \frac{1}{C} \int i \, dt. \tag{2.65}$$

The energy stored in a capacitor depends on the capacitance and the voltage across the capacitor and is given by

$$E = \frac{1}{2} C v_C^2. \tag{2.66}$$

Electrical circuits are modelled using Kirchhoff's laws. There are two laws: Kirchhoff's current law and Kirchhoff's voltage law. To apply these laws effectively, a sign convention should be employed.

Kirchhoff's current law The sum of the currents at a node in a circuit is zero, i.e. the total current flowing into any junction in a circuit is equal to the total current leaving the junction.

Figure 2.14 shows the sign convention that can be employed when using Kirchhoff's current law. We can write

$$i_1 + i_2 + i_3 = 0$$

for the circuit in Figure 2.14(a),

$$-(i_1 + i_2 + i_3) = 0$$

for the circuit in Figure 2.14(b) and

$$i_1 + i_2 - i_3 = 0$$

for the circuit in Figure 2.14(c).

Kirchhoff's voltage law The sum of voltages around any loop in a circuit is zero, i.e. in a circuit containing a source of electromotive force (e.m.f.), the algebraic sum of the potential drops across each circuit element is equal to the algebraic sum of the applied e.m.f.s.

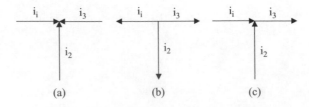

| (a) | (b) | (c) |

Figure 2.14   Applying Kirchhoff's current law

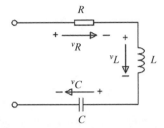

Figure 2.15   Applying Kirchhoff's voltage law

It is important to observe the sign convention when applying Kirchhoff's voltage law. An example circuit is given in Figure 2.15. For this circuit we can write.

$$v_R + v_L + v_C = 0.$$

Some examples of the modelling of electrical circuits are given below.

Example 2.8

Figure 2.16 shows a simple electrical circuit consisting of a resistor, an inductor and a capacitor. A voltage V_a is applied to the circuit. Derive an expression for the mathematical model for this system.

Solution

Applying Kirchhoff's voltage law, we can write

$$v_R + v_L + v_C = V_a$$

or

$$Ri + L\frac{di}{dt} + \frac{1}{C} \int i\,dt = V_a. \tag{2.67}$$

For the capacitor we can write

$$i = C\frac{dv_C}{dt}.$$

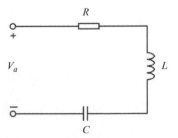

Figure 2.16   Simple electrical circuit

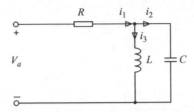

Figure 2.17   Electrical circuit for the Example 2.9

Substituting this into (2.67), we obtain

$$RC\frac{dv_C}{dt} + LC\frac{d^2v_C}{dt^2} + v_C = V_a \tag{2.68}$$

which can also be written as

$$LC\ddot{v}_C + RC\,\dot{v}_C + v_C = V_a. \tag{2.69}$$

Example 2.9

Figure 2.17 shows an electrical circuit consisting of a capacitor, an inductor and a resistor. The inductor and the capacitor are connected in parallel. A voltage V_a is applied to the circuit. Derive a mathematical model for the system.

Solution

Applying Kirchhoff's current law, we can write

$$i_1 = i_2 + i_3. \tag{2.70}$$

Now, the potential difference across the inductor and also across the capacitor is v_C. Similarly, the potential difference across the resistor is $V_a - v_C$. Thus,

$$i_i = \frac{V_a - v_C}{R}, \tag{2.71}$$

$$i_2 = C\frac{dv_C}{dt}, \tag{2.72}$$

$$i_3 = \frac{1}{L}\int v_C dt. \tag{2.73}$$

Combining (2.70)–(2.73) we obtain,

$$\frac{V_a - v_C}{R} = C\frac{dv_C}{dt} + \frac{1}{L}\int v_C dt$$

or

$$\frac{R}{L}\int v_C dt + RC\frac{dv_C}{dt} + v_C = V_a.$$

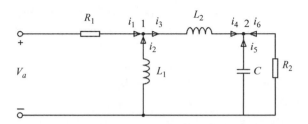

Figure 2.18 Circuit for Example 2.18

Example 2.10

Figure 2.18 shows an electrical circuit consisting of two inductors, two resistors and a capacitor. A voltage V_a is applied to the circuit. Derive an expression for the mathematical model for the circuit.

Solution

The circuit consists of two nodes and two loops. We can apply Kirchhoff's current law to the nodes. For node 1,

$$i_1 + i_2 + i_3 = 0$$

or

$$\frac{V_a - v_1}{R_1} + \frac{1}{L_1} \int (0 - v_1)dt + \frac{1}{L_2} \int (v_2 - v_1)dt = 0. \tag{2.74}$$

Differentiating (2.74) with respect to time, we obtain

$$\frac{\dot{V}_a}{R_1} - \frac{\dot{V}_1}{R_1} - \frac{v_1}{L_1} + \frac{v_2}{L_2} - \frac{v_1}{L_2} = 0$$

or

$$\frac{\dot{V}_a}{R_1} = \frac{\dot{V}_1}{R_1} + \left(\frac{1}{L_1} + \frac{1}{L_2}\right)v_1 - \frac{v_2}{L_2}. \tag{2.75}$$

For node 2,

$$i_4 + i_5 + i_6 = 0$$

or

$$\frac{1}{L_2} \int (v_1 - v_2)dt + C\frac{d(0 - v_2)}{dt} + \frac{0 - v_2}{R_2}. \tag{2.76}$$

Differentiating (2.76) with respect to time, we obtain

$$\frac{v_1 - v_2}{L_2} - C\ddot{v}_2 - \frac{\dot{v}_2}{R_2} = 0$$

which can be written as

$$C\ddot{v}_2 + \frac{\dot{v}_2}{R_2} - \frac{v_1}{L_2} + \frac{v_2}{L_2} = 0. \tag{2.77}$$

Equations (2.75) and (2.76) describe the operation of the circuit. These two equations can be represented in matrix form as

$$\begin{bmatrix} 0 & 0 \\ 0 & C \end{bmatrix}\begin{bmatrix} \ddot{v}_1 \\ \ddot{v}_2 \end{bmatrix} + \begin{bmatrix} 1/R_1 & 0 \\ 0 & 1/R_2 \end{bmatrix}\begin{bmatrix} \dot{v}_1 \\ \dot{v}_2 \end{bmatrix} + \begin{bmatrix} 1/L_1 + 1/L_2 & -1/L_2 \\ -1/L_2 & 1/L_2 \end{bmatrix}\begin{bmatrix} v_1 \\ v_2 \end{bmatrix} = \begin{bmatrix} \dot{V}_a \\ R_1 \\ 0 \end{bmatrix}.$$

2.3 ELECTROMECHANICAL SYSTEMS

Electromechanical systems such as electric motors and electric pumps are used in most industrial and commercial applications. Figure 2.19 shows a simple d.c. motor circuit. The torque produced by the motor is proportional to the applied current and is given by

$$T = k_t i, \tag{2.78}$$

where T is the torque produced, k_t is the torque constant and i is the motor current. Assuming there is no load connected to the motor, the motor torque can be expressed as

$$T = I\frac{d\omega}{dt}$$

or

$$I\frac{d\omega}{dt} = k_t i. \tag{2.79}$$

As the motor armature coil is rotating in a magnetic field there will be a *back e.m.f.* induced in the coil in such a way as to oppose the change producing it. This e.m.f. is proportional to the angular speed of the motor and is given by:

$$v_b = k_e \omega, \tag{2.80}$$

where v_b is the back e.m.f., k_e is the back e.m.f. constant, and ω is the angular speed of the motor.

Using Kirchhoff's voltage law, we can write the following equation for the motor circuit:

$$V_a - v_b = L\frac{di}{dt} + Ri, \tag{2.81}$$

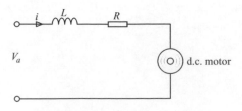

Figure 2.19   Simple d.c. motor

where V_a is the applied voltage, and L and R are the inductance and the resistance of the armature circuit, respectively. From (2.79),

$$i = \frac{I}{k_t}\frac{d\omega}{dt} \tag{2.82}$$

Combining (2.80)–(2.82), we obtain

$$\frac{LI}{k_t}\frac{d^2\omega}{dt^2} + \frac{RI}{k_t}\frac{d\omega}{dt} + k_e\omega = V_a. \tag{2.83}$$

Equation (2.83) is the model for a simple d.c. motor, describing the change of the angular velocity with the applied voltage. In many applications the motor inductance is small and can be neglected. The model then becomes

$$\frac{RI}{k_t}\frac{d\omega}{dt} + k_e\omega = V_a.$$

Models of more complex d.c. motor circuits are given in the following examples.

Example 2.11

Figure 2.20 shows a d.c. motor circuit with a load connected to the motor shaft. Assume that the shaft is rigid, has negligible mass and has no torsional spring effect or rotational damping associated with it. Derive an expression for the mathematical model for the system.

Solution

Since the shaft is assumed to be massless, the moments of inertia of the rotor and the load can be combined into I, where

$$I = I_M + I_L$$

where I_M is the moment of inertia of the motor and I_L is the moment of inertia of the load. Using Kirchhoff's voltage law, we can write the following equation for the motor circuit:

$$V_a - v_b = L\frac{di}{dt} + Ri, \tag{2.84}$$

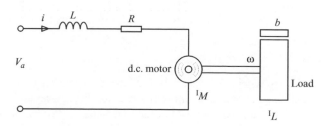

Figure 2.20   Direct current motor circuit for Example 2.11

where V_a is the applied voltage and L and R are the inductance and the resistance of the armature circuit, respectively. Substituting (2.80), we obtain

$$V_a = L\frac{di}{dt} + Ri + k_e\omega$$

or

$$V_a = L\dot{i} + Ri + k_e\dot{\theta}. \tag{2.85}$$

We can also write the torque equation as

$$T + T_L - b\omega = I\frac{d\omega}{dt}.$$

Using (2.78),

$$I\frac{d\omega}{dt} + b\omega - k_t i = T_L$$

or

$$I\ddot{\theta} + b\dot{\theta} - k_t i = T_L. \tag{2.86}$$

Equations (2.85) and (2.86) describe the model of the circuit. These two equations can be represented in matrix form as

$$\begin{bmatrix} I & 0 \\ 0 & 0 \end{bmatrix} \begin{bmatrix} \ddot{\theta} \\ \dot{i} \end{bmatrix} + \begin{bmatrix} b & 0 \\ k_e & L \end{bmatrix} \begin{bmatrix} \dot{\theta} \\ i \end{bmatrix} + \begin{bmatrix} 0 & -k_t \\ 0 & R \end{bmatrix} \begin{bmatrix} \theta \\ i \end{bmatrix} = \begin{bmatrix} T_L \\ V_a \end{bmatrix}.$$

2.4 FLUID SYSTEMS

Gases and liquids are collectively referred to as fluids. Fluid systems are used in many industrial as well as commercial applications. For example, liquid level control is a well-known application of liquid systems. Similarly, gas systems are used in robotics and in industrial movement control applications.

In this section, we shall look at the models of simple liquid systems (or hydraulic systems).

2.4.1 Hydraulic Systems

The basic elements of hydraulic systems are *resistance*, *capacitance* and *inertance* (see Figure 2.21). These elements are similar to their electrical equivalents of resistance, capacitance and inductance. Similarly, electrical current is equivalent to volume flow rate, and the potential difference in electrical circuits is similar to pressure difference in hydraulic systems.

Hydraulic resistance

Hydraulic resistance occurs whenever there is a pressure difference, such as liquid flowing from a pipe of one diameter to to one of a different diameter. If the pressures at either side of a hydraulic resistance are p_1 and p_2, then the hydraulic resistance R is defined as

$$p_1 - p_2 = Rq$$

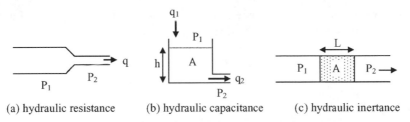

(a) hydraulic resistance   (b) hydraulic capacitance   (c) hydraulic inertance

Figure 2.21   Hydraulic system elements

where q is the volumetric flow rate of the fluid.

Hydraulic capacitance

Hydraulic capacitance is a measure of the energy storage in a hydraulic system. An example of hydraulic capacitance is a tank which stores energy in the form of potential energy. Consider the tank shown in Figure 2.21(b). If q_1 and q_2 are the inflow and outflow, respectively, and V is the volume of the fluid inside the tank, we can write

$$q_1 - q_2 = \frac{dV}{dt} = A\frac{dh}{dt}. \tag{2.87}$$

Now, the pressure difference is given by

$$p_1 - p_2 = h\rho g = p$$

or

$$h = \frac{p}{\rho g}. \tag{2.88}$$

Substituting in (2.87), we obtain

$$q_1 - q_2 = \frac{A}{\rho g}\frac{dp}{dt}. \tag{2.89}$$

Writing (2.89) as

$$q_1 - q_2 = C\frac{dp}{dt}, \tag{2.90}$$

we then arrive at the definition of hydraulic capacitance:

$$C = \frac{A}{\rho g}. \tag{2.91}$$

Note that (2.90) is similar to the expression for a capacitor and can be written as

$$p = \frac{1}{C}\int (q_1 - q_2)dt. \tag{2.92}$$

Hydraulic inertance

Hydraulic inertance is similar to the inductance in electrical systems and is derived from the inertia force required to accelerate fluid in a pipe.

Let $p_1 - p_2$ be the pressure drop that we want to accelerate in a cross-sectional area of A, where m is the fluid mass and v is the fluid velocity. Applying Newton's second law, we can write

$$m\frac{dv}{dt} = A(p_1 - p_2).$$

(2.93)

If the pipe length is L, then the mass is given by

$$m = L\rho A.$$

We can now write (2.93) as

$$L\rho A\frac{dv}{dt} = A(p_1 - p_2)$$

or

$$p_1 - p_2 = L\rho\frac{dv}{dt},$$

(2.94)

but the rate of flow is given by $q = Av$, so (2.94) can be written as

$$p_1 - p_2 = \frac{L\rho}{A}\frac{dq}{dt}.$$

(2.95)

The inertance I is then defined as

$$I = \frac{L\rho}{A},$$

and thus the relationship between the pressure difference and the flow rate is similar to the relationship between the potential difference and the current flow in an inductor, i.e.

$$p_1 - p_2 = I\frac{dq}{dt}.$$

(2.96)

Models of some hydraulic systems are given below.

Example 2.12

Figure 2.22 shows a liquid level system where liquid enters a tank at the rate of q_i and leaves at the rate of q_o through an orifice. Derive the mathematical model for the system, showing the relationship between the height h of the liquid and the input flow rate q_i.

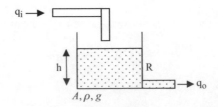

Figure 2.22   Liquid level system

Solution

From (2.89),

$$q_i - q_o = \frac{A}{\rho g}\frac{dp}{dt}$$

or

$$q_i = \frac{A}{\rho g}\frac{dp}{dt} + q_o. \qquad (2.97)$$

Recalling that

$$p = h\rho g,$$

(2.97) becomes

$$q_i = A\frac{dh}{dt} + q_o. \qquad (2.98)$$

Since

$$p_1 - p_2 = Rq_o,$$

so that

$$q_o = \frac{p_1 - p_2}{R} = \frac{h\rho g}{R},$$

substituting in (2.98) gives

$$q_i = A\frac{dh}{dt} + \frac{\rho g}{R}h. \qquad (2.99)$$

Equation (2.99) shows the variation of the height of the water with the inflow rate. If we take the Laplace transform of both sides, we obtain

$$q_i(s) = Ash(s) + \frac{\rho g}{R}h(s)$$

and the transfer function of the system can be written as

$$\frac{h(s)}{q_i(s)} = \frac{1}{As + \rho g/R};$$

the block diagram is shown in Figure 2.23.

Example 2.13

Figure 2.24 shows a two-tank liquid level system where liquid enters the first tank at the rate of q_i and then flows to the second tank at the rate of q_1 through an orifice R_1. Water then leaves

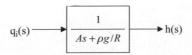

Figure 2.23   Block diagram of the liquid level system

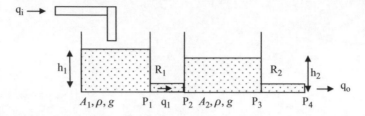

Figure 2.24   Two tank liquid level system

the second tank at the rate of q_o through an orifice of R_2. Derive the mathematical model for the system.

Solution

The solution is similar to Example 2.12, but we have to consider both tanks.
  For tank 1,

$$q_i - q_1 = \frac{A_1}{\rho g}\frac{dp}{dt}$$

or

$$q_i = \frac{A_1}{\rho g}\frac{dp}{dt} + q_1. \tag{2.100}$$

But

$$p = h\rho g,$$

thus (2.100) becomes

$$q_i = A_1\frac{dh_1}{dt} + q_1 \tag{2.101}$$

Since

$$p_1 - p_2 = R_1 q_1$$

or

$$q_1 = \frac{p_1 - p_2}{R_1} = \frac{h_1\rho g - h_2\rho g}{R_1},$$

we have

$$q_i = A_1\frac{dh_1}{dt} + \frac{\rho g h_1}{R_1} - \frac{\rho g h_2}{R_1}. \tag{2.102}$$

For tank 2,

$$q_1 - q_0 = \frac{A_2}{\rho g}\frac{dp}{dt}, \tag{2.103}$$

and with

$$p = h\rho g$$

(2.103) becomes

$$q_1 - q_o = A_2 \frac{dh_2}{dt}. \tag{2.104}$$

But

$$q_1 = \frac{p_1 - p_2}{R_1} \quad \text{and} \quad q_o = \frac{p_2 - p_3}{R_2}$$

so

$$q_1 - q_0 = \frac{p_1 - p_2}{R1} - \frac{p_2 - p_3}{R_2} = \frac{h_1 \rho g - h_2 \rho g}{R_1} - \frac{h_2 \rho g}{R_2}.$$

Substituting in (2.104), we obtain

$$A_2 \frac{dh_2}{dt} - \frac{\rho g h_1}{R_1} + \left(\frac{1}{R_1} + \frac{1}{R_2} \right) \rho g h_2 = 0. \tag{2.105}$$

Equations (2.102) and (2.105) describe the behaviour of the system. These two equations can be represented in matrix form as

$$\begin{bmatrix} A_1 & 0 \\ 0 & A_2 \end{bmatrix} \begin{bmatrix} \dot{h}_1 \\ \dot{h}_2 \end{bmatrix} + \begin{bmatrix} \rho g / R_1 & -\rho g / R_1 \\ -\rho g / R_1 & \rho g / R_1 + \rho g / R_2 \end{bmatrix} \begin{bmatrix} h_1 \\ h_2 \end{bmatrix} = \begin{bmatrix} q_i \\ 0 \end{bmatrix}.$$

2.5 THERMAL SYSTEMS

Thermal systems are encountered in chemical processes, heating, cooling and air conditioning systems, power plants, etc. Thermal systems have two basic components: thermal resistance and thermal capacitance. Thermal resistance is similar to the resistance in electrical circuits. Similarly, thermal capacitance is similar to the capacitance in electrical circuits. The across variable, which is measured across an element, is the temperature, and the through variable is the heat flow rate. In thermal systems there is no concept of inductance or inertance. Also, the product of the across variable and the through variable is not equal to power. The mathematical modelling of thermal systems is usually complex because of the complex distribution of the temperature. Simple approximate models can, however, be derived for the systems commonly used in practice.

Thermal resistance, R, is the resistance offered to the heat flow, and is defined as:

$$R = \frac{T_2 - T_1}{q}, \tag{2.106}$$

where T_1 and T_2 are the temperatures, and q is the heat flow rate.

Thermal capacitance is a measure of the energy storage in a thermal system. If q_1 is the heat flowing into a body and q_2 is the heat flowing out then the difference $q_1 - q_2$ is stored by the body, and we can write

$$q_1 - q_2 = mc \frac{dT}{dt}, \tag{2.107}$$

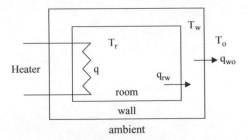

Figure 2.25 Simple thermal system

where m is the mass and c is the specific heat capacity of the body. If we let the heat capacity be denoted by C, then

$$q_1 - q_2 = C\frac{dT}{dt}, \qquad (2.108)$$

where $C = mc$.

An example thermal system model is given below.

Example 2.14

Figure 2.25 shows a room heated with an electric heater. The inside of the room is at temperature T_r and the walls are assumed to be at temperature T_w. If the outside temperature is T_o, develop a model of the system to show the relationship between the supplied heat q and the room temperature T_r.

Solution

The heat flow from inside the room to the walls is given by

$$q_{rw} = \frac{T_r - T_w}{R_r}, \qquad (2.109)$$

where R_r is the thermal resistance of the room.

Similarly, the heat flow from the walls to the outside is given by

$$q_{wo} = \frac{T_w - T_o}{R_w}, \qquad (2.110)$$

where R_w is the thermal resistance of the walls.

Using (2.108) and (2.109), we can write

$$q - \left(\frac{T_r - T_w}{R_r}\right) = C_1\frac{dT_r}{dt}$$

or

$$C_1\dot{T}_r + \frac{T_r}{R_r} - \frac{T_w}{R_r} = q. \qquad (2.111)$$

Also, using (2.108) and (2.110), we can write

$$\left(\frac{T_r - T_w}{R_r}\right) - \left(\frac{T_w - T_o}{R_w}\right) = C_2 \frac{dT_w}{dt}$$

or

$$C_2 \dot{T}_w - \frac{T_r}{R_r} + \left(\frac{1}{R_r} + \frac{1}{R_w}\right) T_w = \frac{T_o}{R_w}. \tag{2.112}$$

Equations (2.111) and (2.112) describe the behaviour of the system and they can be written in matrix form as

$$\begin{bmatrix} C_1 & 0 \\ 0 & C_2 \end{bmatrix} \begin{bmatrix} \dot{T}_r \\ \dot{T}_w \end{bmatrix} + \begin{bmatrix} 1/R_r & -1/R_r \\ -1/R_r & 1/R_r + 1/R_w \end{bmatrix} \begin{bmatrix} T_r \\ T_w \end{bmatrix} = \begin{bmatrix} q \\ T_o/R_w \end{bmatrix}.$$

Example 2.15

Figure 2.26 shows a heated stirred tank thermal system. Liquid enters the tank at the temperature T_i with a flow rate of W. The water is heated inside the tank to temperature T. The temperature leaves the tank at the same flow rate of W. Derive a mathematical model for the system, assuming that there is no heat loss from the tank.

Solution

The following equation can be written for the conservation of energy:

$$Q_p + Q_i = Q_l + Q_o, \tag{2.113}$$

where Q_p is the heat supplied by the heater, Q_i is the heat flow via the liquid entering the tank, Q_l is the heat flow into the liquid and Q_o is the heat flow via the liquid leaving the tank.
   Now,

$$Q_i = WC_pT_i \tag{2.114}$$

where W is the flow rate (kg/s), and C_p is the specific heat capacity of the liquid. Also

$$Q_o = WC_pT \tag{2.115}$$

and

$$Q_l = C\frac{dT}{dt}, \tag{2.116}$$

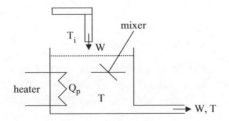

Figure 2.26   Heated stirred tank for Example 2.15

where C is the thermal capacity, i.e. $C = \rho V C_p$ and V is the volume of the tank. Substituting (2.114)–(2.116) into (2.113) gives

$$Q_p + W C_p T_i = C \frac{dT}{dt} + W C_p T$$

or

$$\frac{dT}{dt} = \frac{W C_p (T_i - T) + Q_p}{\rho V C_p}.$$

2.6 EXERCISES

1. Figure 2.27 shows a simple mechanical system consisting of a mass, spring and damper. Derive a mathematical model for the system, determine the transfer function, and draw the block diagram.

2. Consider the system of two massless springs shown in Figure 2.28. Derive a mathematical model for the system.

3. Three massless springs with the same stiffness constant are connected in series. Derive an expression for the equivalent spring stiffness constant.

4. Figure 2.29 shows a simple mechanical system. Derive an expression for the mathematical model for the system.

5. Figure 2.30 shows a rotational mechanical system. Derive an expression for the mathematical model for the system.

6. Two rotational springs are connected in parallel. Derive an expression for the equivalent spring stiffness constant.

7. Figure 2.31 shows a simple system with a gear-train. Derive an expression for the mathematical model for the system.

8. A simple electrical circuit is shown in Figure 2.32. Derive an expression for the mathematical model for the system.

9. Figure 2.33 shows an electrical circuit. Use Kirchhoff's laws to derive the mathematical model for the system.

10. A liquid level system is shown in Figure 2.34, where q_i and q_o are the inflow and outflow rates, respectively. The system has two fluid resistances, R_1 and R_2, in series. Derive an expression for the mathematical model for the system.

11. Figure 2.35 shows a liquid level system with three tanks. Liquid enters the first tank at the rate q_i and leaves the third tank at the rate q_o. Assume that all tanks have the same dimensions. Derive an expression for the mathematical model for this system.

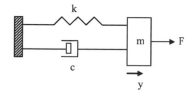

Figure 2.27   Simple mechanical system for Exercise 1

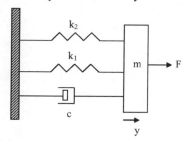

Figure 2.28   System of two massless springs for Exercise 2

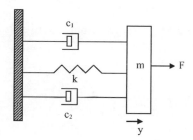

Figure 2.29   Simple mechanical system for Exercise 4

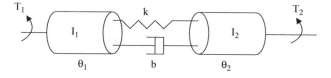

Figure 2.30   Simple mechanical system for Exercise 5

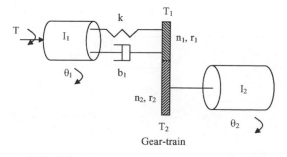

Figure 2.31   Simple system with a gear-train for Exercise 7

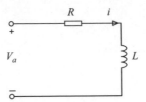

Figure 2.32   Simple electrical circuit for Exercise 8

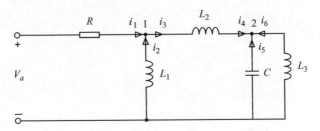

Figure 2.33   Electrical circuit for Exercise 9

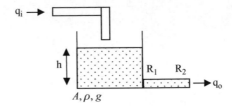

Figure 2.34   Liquid level system for Exercise 10

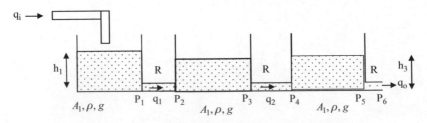

Figure 2.35   Three-tank liquid level system for Exercise 11

FURTHER READING

[Cannon, 1967]   Cannon, R.H. Jr. Dynamics of Physical Systems. McGraw-Hill, New York, 1967.

[Cochin, 1980]   Cochin, I. Analysis and Design of Dynamic Systems. Harper & Row, New York, 1980.

[D'Souza, 1988]   D'Souza, A. Design of Control Systems. Prentice Hall, Englewood Cliffs, NJ, 1988.

[Franklin and Powell, 1986] Franklin, G.F. and Powell, J.D. Feedback Control of Dynamic Systems. Addison-Wesley, Reading, MA, 1986.

[Leigh, 1985]   Leigh, J.R. Applied Digital Control. Prentice Hall, Englewood Cliffs, NJ, 1985.

[Ogata, 1990]   Ogata, K. Modern Control Engineering 2nd edn., Prentice Hall, Englewood Cliffs, NJ, 1990.

3

The PIC Microcontroller

3.1 THE PIC MICROCONTROLLER FAMILY

The PIC microcontroller is a family of microcontrollers manufactured by the Microchip Technology Inc. Currently the PIC is one of the most popular microcontrollers used in education, and in commercial and industrial applications. The family consists of over 140 devices, ranging from simple 4-pin dual in-line devices with 0.5 K memory, to 80-pin complex devices with 32 K memory.

Even though the family consists of a large number of devices, all the devices have the same basic structure, offering the following fundamental features:

- reduced instruction set (RISC) with only 35 instructions;
- bidirectional digital I/O ports;
- RAM data memory;
- rewritable flash, or one-time programmable program memory;
- on-chip timer with pre-scaler;
- watchdog timer;
- power-on reset;
- external crystal operation;
- 25 mA current source/sink capability;
- power-saving sleep mode.

More complex devices offer the following additional features:

- analog input channels;
- analog comparators;
- serial USART;
- nonvolatile EEPROM memory;
- additional on-chip timers;
- external and internal (timer) interrupts;
- PWM output;
- CAN bus interface;
- I^2C bus interface;
- USB interface;
- LCD interface.

Microcontroller Based Applied Digital Control   D. Ibrahim
© 2006 John Wiley & Sons, Ltd

The PIC microcontroller product family currently consists of six groups:

- PIC10FXXX 12-bit program word;
- PIC12CXXX/PIC12FXXX 12/14-bit program memory;
- PIC16C5X 12-bit program word;
- PIC16CXXX/PIC16FXXX 14-bit program word;
- PIC17CXXX 16-bit program word;
- PIC18CXXX/PIC18FXXX 16-bit program word.

3.1.1   The 10FXXX Family

Table 3.1 gives a summary of the features of this family. PIC10F200 is a member of this family with the following features.

PIC10F200. This microcontroller is available in a 6-pin SOT-23 package (see Figure 3.1), or in 8-pin PDIP package (see Figure 3.2). The device has 33 instructions, 256×12 word flash program memory, 16 bytes of RAM data memory, four I/O ports, and one 8-bit timer. Clocking is from a precision 4 MHz internal oscillator. Other members of the family have larger memories and also an internal comparator.

Table 3.1   Some PIC10FXXX family members

Microcontroller	Program memory	Data RAM	I/O pins	Comparators	A/D converters
10F200	256×12	16	4	0	0
10F202	512×12	24	4	0	0
10F204	256×12	16	4	1	0
10F206	512×12	24	4	1	0
10F220	256×12	16	4	0	3/8-bit

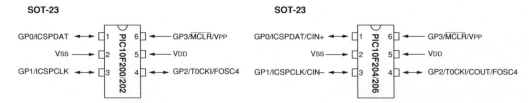

Figure 3.1   PIC10FXXX family in 6-pin SOT-23 package

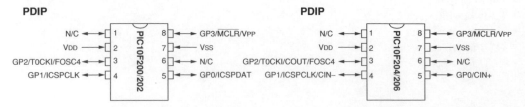

Figure 3.2   PIC10FXXX family in 8-pin PDIP package

Table 3.2   Some PIC12CXXX family members

Microcontroller	Program memory	Data RAM	I/O pins	EEPROM	A/D converters
12C508	256 × 12	25	6	0	0
12C509	1024 × 12	41	6	0	0
12C671	1024 × 14	128	6	0	4
12F629	1024 × 14	64	6	128	0
12F675	1024 × 14	64	6	128	4

3.1.2 The 12CXXX/PIC12FXXX Family

Table 3.2 gives a summary of the features of this family. The PIC12C508 is a member of this family with the following features.

PIC12C508. This is another low-cost microcontroller available in an 8-pin dual in-line package. The device has 512 × 12 word flash memory, 25 bytes of RAM data memory, six I/O ports, and one 8-bit timer. Operation is from a 4 MHz clock. Other members of the family have larger memories, higher speed, and A/D converters (e.g. 12C672).

The PIC12FXXX family has the same structure as the 12CXXX but with flash program memory and additional EEPROM data memory.

3.1.3 The 16C5X Family

Table 3.3 gives a summary of the features of this family. These devices have 14-, 18-, 20- and 28-pin packages. The PIC16C54 is a member of this family with the following features.

PIC16C54. This is an 18-pin microcontroller with 384 × 12 EPROM program memory. The device has 25 bytes of RAM, 12 I/O port pins, a timer and a watchdog timer. Other members of the family have larger memories and more I/O ports.

3.1.4 The 16CXXX Family

Table 3.4 gives a summary of the features of this family. These devices are similar to the 16CXX series, but they have 14 bits of program memory and some of them have A/D converters. The PIC16C554 is a member of this family with the following features.

Table 3.3   Some PIC16C5X family members

Microcontroller	Program memory	Data RAM	I/O pins	EEPROM	A/D converters
16C54	512 × 12	25	12	0	0
16C56	1024 × 12	25	12	0	0
16C58	2048 × 12	73	12	0	0
16C505	1024 × 12	72	12	0	0

Table 3.4   Some PIC16CXXX family members

Microcontroller	Program memory	Data RAM	I/O pins	EEPROM	A/D converters
16C554	512 × 14	80	13	0	0
16C620	512 × 14	96	13	0	0
16C642	4096 × 14	176	22	0	0
16C716	2048 × 14	128	13	0	4
16C926	8192 × 14	336	52	0	5

PIC16C554. This is an 18-pin device with 512 × 14 program memory. The data memory is 80 bytes and 13 I/O port pins are provided. Other members of this family provide A/D converters (e.g. PIC16C76), USART capability (e.g. PIC16C67), larger memory, more I/O port pins and higher speed.

The PIC16FXXX family (e.g. PIC16F74) is upward compatible with the PIC16CXXX family. These devices are also 14 bits wide and have flash program memory and an internal 4 MHz clock oscillator as added features.

3.1.5 The 17CXXX Family

These are 16-bit microcontrollers. The program memory capacity ranges from 8192 × 16 (e.g. PIC17C42) to 16 384 × 16 (e.g. PIC17C766). The devices also have larger RAM data memories, higher current sink capabilities and larger I/O port pins, e.g. the PIC17C766 provides 66 I/O port pins. Table 3.5 gives a summary of the features of this family.

3.1.6 The PIC18CXXX Family

These are high-speed 16-bit microcontrollers, with maximum clock frequency 40 MHz. The devices in this family have large program and data memories, a large number of I/O pins, and A/D converters. They have an instruction set with 77 instructions, including multiplication. Table 3.6 gives a summary of the features of this family.

PIC18FXXX family members are upward compatible with PIC18CXXX. These microcontrollers in addition offer flash program memories and EEPROM data memories. Some members of the family provide up to 65 536 × 16 program memories and 3840 bytes of RAM memory.

Table 3.5   Some PIC17CXXX family members

Microcontroller	Program memory	Data RAM	I/O pins	EEPROM	A/D converters
17C42	2048 × 16	232	33	0	0
17C43	4096 × 16	454	33	0	0
17C762	8192 × 16	678	66	0	16
17C766	16384 × 16	902	66	0	16

Table 3.6   Some PIC18CXXX family members

Microcontroller	Program memory	Data RAM	I/O pins	EEPROM	A/D converters
18C242	8192 × 16	512	2	0	5
18C452	16384 × 16	1536	34	0	8
18C658	16384 × 16	1536	52	0	12
18C858	16384 × 16	1536	68	0	16
18F242	8192 × 16	256	23	256	5

3.2 MINIMUM PIC CONFIGURATION

The minimum PIC configuration depends on the type of microcontroller used. Normally, the operation of a PIC microcontroller requires a power supply, reset circuit and oscillator.

The power supply is usually +5 V and, as shown in Figure 3.3, can be obtained from the mains supply by using a step-down transformer, a rectifier circuit and a power regulator chip, such as the LM78L05.

Although PIC microcontrollers have built-in power-on reset circuits, it is useful in many applications to have external reset circuits. When the microcontroller is reset, all of its special function registers are put into a known state and execution of the user program starts from address 0 of the program memory.

As shown in Figure 3.4, reset is normally achieved by connecting a 4.7 K pull-up resistor from the master clear (MCLR) input to the supply voltage. Sometimes the voltage rises too slowly and the simple reset circuit may not work. In this case, the circuit shown in Figure 3.5 is recommended.

In many applications it may be required to reset the microcontroller by pressing an external button. The circuit given in Figure 3.6 enables the microcontroller to reset when the button is pressed.

PIC microcontrollers have built-in clock oscillator circuits. Additional components are needed to enable such clock oscillator circuits to function; some PIC microcontrollers have these built in, while others require external components. The internal oscillator can be operated in one of six modes:

- external oscillator;
- LP – low-power crystal;
- XT – low-speed crystal/resonator;

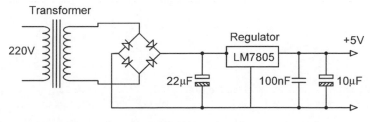

Figure 3.3   A simple microcontroller power source

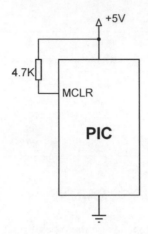

Figure 3.4   Simple reset circuit

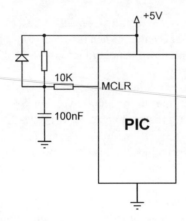

Figure 3.5   Recommended reset circuit

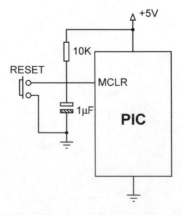

Figure 3.6   Push-button reset circuit

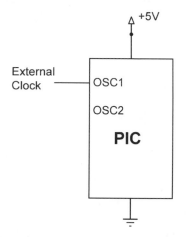

Figure 3.7   Using an external oscillator

* HS – high-speed crystal/resonator;
* RC – resistor and capacitor;
* no external components (only some PICs).

3.2.1 External Oscillator

An external oscillator can be connected to the OSC1 input as shown in Figure 3.7. The oscillator should generate square wave pulses at the required frequency. The timing accuracy depends upon the accuracy of this external oscillator. When operated in this mode, the chip should be programmed for LP, XT or HS clock mode.

3.2.2 Crystal Operation

An external crystal should be used when very accurate timing is required. As shown in Figure 3.8, the crystal should be connected between the OSC1 and OSC2 inputs together with a pair of capacitors. The value of the capacitors should be chosen as in Table 3.7. For example, with a crystal of 4 MHz, two 22 pF capacitors can be used.

3.2.3 Resonator Operation

Resonators are usually available in the frequency range of about 4–8 MHz. Although resonators are not as accurate as crystals, they are usually accurate enough for most applications. Resonators have the advantages that they are low-cost and only one component is required compared to three components in the case of crystals (the crystal itself and two capacitors). Figure 3.9 shows how a resonator can be used with PIC microcontrollers.

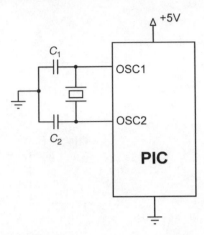

Figure 3.8   Using a crystal

Table 3.7   Capacitor values for crystal operation

Mode	Frequency	C1, C2 (pF)
LP	32 kHz	68–100
LP	200 kHz	15–33
XT	100 kHz	100–150
XT	2 MHz	15–33
XT	4 MHz	15–33
HS	4 MHz	15–33
HS	10 MHz	15–33

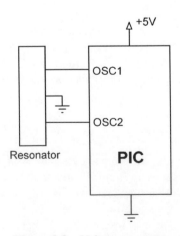

Figure 3.9   Using a resonator

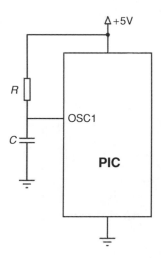

Figure 3.10 Using an RC circuit

3.2.4 RC Operation

There are many low-cost applications where the timing accuracy is not important – flashing an LED every second, scanning a keyboard, reading the temperature every second, etc. In such applications the clock pulses can be generated by using an external resistor and a capacitor. As shown in Figure 3.10, the resistor and the capacitor should be connected to the OSC1 input of the microcontroller.

The oscillator frequency depends upon the values of the resistor and the capacitor, the supply voltage and environmental factors, such as the temperature. Table 3.8 gives a list of typical resistor and capacitor values for most of the frequencies of interest. For example, a 5 K resistor and a 20 pF capacitor can be used to generate a clock frequency of about 4 MHz.

3.2.5 Internal Clock

Some PIC microcontrollers (e.g. PIC12C672) have built-in clock generation circuitry and do not require any external components to generate the clock pulses. The built-in oscillator is

Table 3.8 Resistor and capacitor values

C (pF)	R (kΩ)	Frequency
20	5	4.61 MHz
	10	2.66 MHz
	100	311 kHz
100	5	1.34 MHz
	10	756 kHz
	100	82.8 kHz
300	5	428 kHz
	10	243 kHz
	100	26.2 kHz

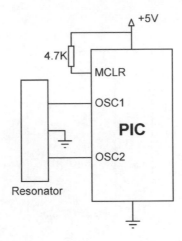

Figure 3.11 Resonator based minimum PIC microcontroller system

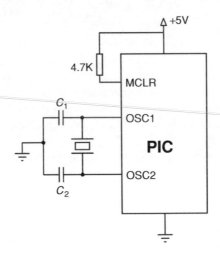

Figure 3.12 Crystal based minimum PIC microcontroller system

usually 4 MHz and can be selected during the programming of the devices. Figure 3.11 shows the circuit diagram of a minimum PIC microcontroller system using a 4 MHz resonator. A minimum PIC microcontroller system using a crystal is shown in Figure 3.12.

3.3 SOME POPULAR PIC MICROCONTROLLERS

In this section we shall look at the architectures of some of the popular PIC microcontrollers in greater detail. We have chosen the popular 18-pin PIC16F84 and the 40-pin PIC16F877 microcontrollers. The architectures and instruction sets of most of the other PIC microcontrollers are very similar, and with the knowledge gained here we should be able to use and program any other PIC microcontroller easily. Since our aim is to program the microcontrollers using a high-level language such as the C, there is no need to learn their exact architecture or assembly instruction set. We shall only look at the features which may be required while developing software using the C programming language.

3.3.1 PIC16F84 Microcontroller

The PIC16F84 is one of the most popular PIC microcontrollers used in many commercial, industrial and hobby applications. This is an 18-pin device which can operate at up to 20 MHz clock speed. It offers 1024×14 flash program memory, 68 bytes of RAM data memory, 64 bytes of EEPROM nonvolatile data memory, 8-bit timer with pre-scaler, watchdog timer, 13 I/O pins, external and internal interrupt sources, and large current sink and source capability.

Figure 3.13 shows the pin configuration of the PIC16F84. The functions of various pins are as follows:

RB0–RB7	Bidirectional port B pins
RA0–RA4	Bidirectional port A pins
Vdd	Supply voltage
Vss	Ground
OSC1	Crystal, resonator, or external clock input
OSC2	Crystal or resonator input
MCLR	Reset input
INT	External interrupt input (shared with RB0)
T0CK1	Optional timer clock input (shared with RA3)

Note that some pin names have a bar on them – for example, $\overline{\text{MCLR}}$ in Figure 3.13. This means that the pin will be active when the applied signal is at logic low (logic 0 in this case).

The PIC16F84 provides four external or internal interrupt sources:

- external interrupt on INT pin;
- timer overflow interrupt;
- state change interrupt on the four higher bits of port B (RB4–RB7);
- EEPROM memory data write complete interrupt.

The data RAM is also known as the *register file map* (RFM) and consists of 80 bytes. As shown in Figure 3.14, the RFM is divided into two parts: the *special function registers* (SFR), and the *general purpose registers* (GPR). The RFM is organized as two banks (more complex PIC microcontrollers may have more banks): bank 0 and bank 1. The bank of a register must

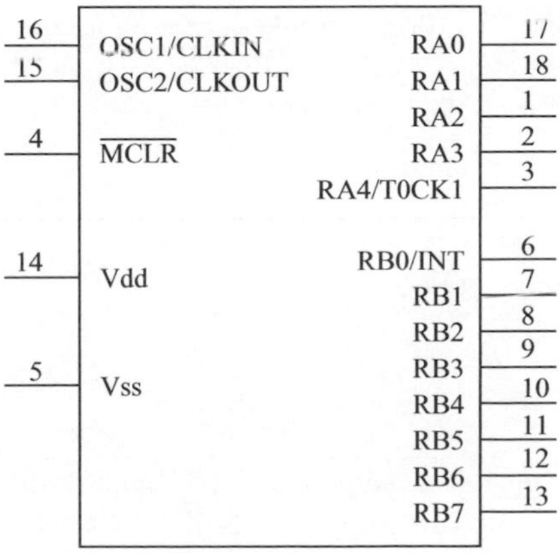

Figure 3.13 PIC16F84 pin configuration

00	Indirect address	Indirect address	80
01	TMR0	OPTION	81
02	PCL	PCL	82
03	STATUS	STATUS	83
04	FSR	FSR	84
05	PORTA	TRISA	85
06	PORTB	TRISB	86
07	–	–	87
08	EEDATA	EECON1	88
09	EEADR	EECON2	89
0A	PCLATH	PCLATH	8A
0B	INTCON	INTCON	8B
0C		**Access in bank 0 (mapped)**	8C
...	**68-byte GPR**		...
...			...
...			...
4F			CF
50	**50 to 7F not used**		D0
...			...
7F			FF
	Bank 0	**Bank 1**	

Figure 3.14   Register file map of PIC16F84

be selected before a register in a bank can be read or written to. Some of the registers are common to both banks.

The SFR are a collection of registers used by the CPU to control the internal operations and the peripherals – setting the I/O direction of a register, sending data to an I/O port, loading the timer register, etc. The SFR used while programming the microcontroller using a high-level language are described in the following sections.

3.3.1.1 OPTION_REG Register

The OPTION_REG register is a readable and writable register at address 0×81 (hexadecimal) of the RFM. This register controls the timer pre-scaler, timer clock edge selection, timer clock source, external interrupt edge selection, and port B pull-up resistors. OPTION_REG bit definitions are given in Figure 3.15. For example, to configure the external interrupt INT pin so that external interrupts are accepted on the falling edge of the INT input, the following bit pattern should be loaded into the OPTION_REG:

X0XXXXXX

where X is a don't-care bit and can be a 0 or a 1.

3.3.1.2 INTCON Register

This is the interrupt control register at addresses $0 \times 0B$ and $0 \times 8B$ of the RFM. The bit definitions of this register are shown in Figure 3.16. INTCON is used to enable/disable the various interrupt sources and interrupt flags. For an interrupt to be accepted by the CPU, the

7	6	5	4	3	2	1	0
RBPU	INTEDG	T0CS	TOSE	PSA	PS2	PS1	PS0

Bit 7: PORTB pull-up control
     1: PORTB pull-ups disabled
     0: PORTB pull-ups enabled

Bit 6: INT external interrupt edge detect
     1: Interrupt on rising edge of INT input
     0: Interrupt on falling edge of INT input

Bit 5: TMR0 timer clock source
     1: T0CK1 external pulse
     0: internal clock

Bit 4: TMR0 source edge select
     1: Increment on HIGH to LOW of T0CK1
     0: Increment on LOW to HIGH of T0CK1

Bit 3: Pre-scaler assignment
     1: Pre-scaler assigned to watchdog timer
     0: Pre-scaler assigned to TMR0

Bit 2-0: Pre-scaler rate
     000   1:2
     001   1:4
     010   1:8
     011   1:16
     100   1:32
     101   1:64
     110   1:128
     111   1:256

Figure 3.15 OPTION_REG bit definitions

7	6	5	4	3	2	1	0
GIE	EEIE	T0IE	INTE	RBIE	T0IF	INTF	RBIF

Bit 7: Global interrupt control
     1: Enable all unmasked interrupts
     0: Disable all interrupts

Bit 6: EEPROM write complete interrupt
     1: Enable EEPROM write complete interrupt
     0: Disable EEPROM write complete interrupt

     Bit 5: TMR0 overflow interrupt
     1: Enable TMR0 interrupt
     0: Disable TMR0 interrupt

Bit 4: INT external interrupt control
     1: Enable INT External interrupt
     0: Disable INT External Interrupt

Bit 3: RB4–RB7 port change interrupt control
     1: Enable RB4–RB7 port change interrupt
     0: Disable RB4–RB7 port change interrupt

Bit 1: INT interrupt flag
     1: INT interrupt occurred
     0: INT interrupt did not occur

Bit 0: RB4–RB7 port change interrupt flag
     1: One or more of RB4–RB7 pins changed state
     0: None of RB4–RB7 pins changed state

Figure 3.16 INTCON bit definitions

following conditions must be met:

- The global interrupt flag in INTCON must be enabled (GIE = 1).

- The interrupt flag of the interrupting source in INTCON must be enabled (e.g. INTE = 1 to enable INT external interrupts).

- Interrupt must physically occur (e.g. INT pin is raised to logic 1 if INTEDG was previously set to logic 1).

After an interrupt is detected the program jumps to the interrupt service routine which is at address 4 of the program memory. At this point further interrupts are disabled and the interrupt flag of the interrupt source (e.g. bit INTF of INTCON for external interrupts) must be cleared for a new interrupt to be accepted from the interrupting source.

3.3.1.3 TRISA and Port A Registers

Port A is a 5-bit-wide port with pins RA0–RA4, and at address 5 of the RFM. Four low-order bits (RA0–RA3) have CMOS output drivers with 25 mA current sink and source capabilities. RA4 is an open-drain port and a suitable pull-up resistor must be connected when used as an output port. Port A pins are bidirectional and the direction of a pin is determined by the settings of register TRISA. Setting a bit in TRISA makes the corresponding port A pin an input. Similarly, clearing a bit in TRISA makes the corresponding port A pin an output. For example, to make bits 0, 1 and 2 of port A input and the other bits output, we have to load TRISA register with:

00000111

3.3.1.4 TRISB and Port B Registers

Port B is a 8-bit-wide port with pins RB0–RB7, and at address 6 of the RFM. The pins have CMOS output drivers with 25 mA current sink and source capabilities. Pin RB0 can be used as an external interrupt pin. Similarly, pins RB4–RB7 can be used to generate an interrupt when the state of any of these pins changes. Port B pins are bidirectional, and the direction of a pin is determined by the settings of register TRISB. Setting a bit in TRISB makes the corresponding port B pin an input. Similarly, clearing a bit in TRISB makes the corresponding port B pin an output. For example, to make bits 0, 2 and 4 of port B input and the other bits output, we have to load TRISB register with:

00010101

3.3.1.5 TMR0 Register

The PIC16F84 provides an 8-bit timer, called TMR0, which can be used either as a timer or a counter. The structure of this timer is shown in Figure 3.17. When used as a counter, the register increments each time a clock pulse is applied to external pin T0CK1 of the microcontroller.

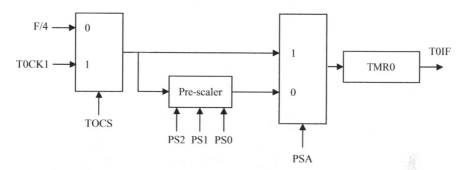

Figure 3.17   TMR0 structure

When used as a timer, the register increments at a rate determined by the microcontroller clock frequency and a pre-scaler, selected by register OPTION_REG. The pre-scaler values range from $1:2$ to $1:256$. For example, when using a 4 MHz clock, the basic instruction cycle is 1 μs (a 4 MHz clock has a period of 0.25 μs, but the clock is internally divided by 4 to obtain the basic instruction cycle). If we select a pre-scaler rate of $1:8$, the timer register will be incremented at every 8 μs.

A timer overflow interrupt is generated when the timer register overflows from 255 to 0. This interrupt can be enabled in software by setting bit 5 of the INTCON register. For example, if we wish to generate interrupts at 200 μs intervals with a 4 MHz clock, we can select a pre-scaler value of $1:4$ and enable timer interrupts. The effective timer clock rate will then be 4 μs. For a time-out of 200 μs, we have to send $200/4 = 50$ clock pulses to the timer register. Thus, the timer register TMR0 should be loaded with $256 - 50 = 206$, i.e. a count of 50 before a timer overflow occurs.

The PIC16F84 microcontroller contains a 64-byte nonvolatile EEPROM memory, controlled by registers EEDATA, EEADR, EECON1 and EECON2. There are instructions to read and write the contents of this memory. EEPROM memory is usually used to store configuration data or maximum and minimum data obtained in real-time measurements.

The PIC16F84 microcontroller also contains a *configuration register* whose bits can be set or cleared during the programming of the device. This register contains bits to select the oscillator mode, to enable or disable code protection, to enable or disable the power-on timer, and to enable or disable the watchdog timer.

3.3.2   PIC16F877 Microcontroller

The PIC16F877 is a 40-pin popular PIC microcontroller. The device offers the following features:

- 8192 × 14 words flash program memory;
- 256 × 8 bytes of EEPROM data memory;
- 368 × 8 RAM data memory;
- eight 10-bit A/D channels;
- 33 bidirectional I/O pins;
- two 8-bit and one 16-bit timers;

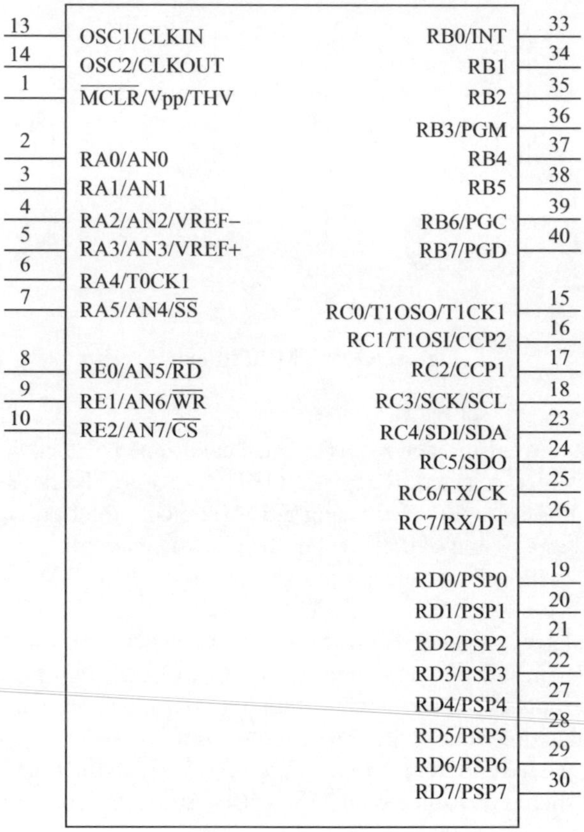

Figure 3.18   PIC16F877 pin configuration

- watchdog timer;
- 14 interrupt sources;
- capture, compare and PWM modules;
- on-chip USART;
- 25 mA current source and sink capability.

Figure 3.18 shows the pin configuration of the PIC16F877. I/O ports are accessed as in the PIC16F84 where each port has a direction register (TRIS) which determines the mode of the I/O pins. One of the nice features of the PIC16F877 is that it contains a multiplexed eight-channel A/D converter with 10-bit resolution. A/D conversion is important in microcontroller based control applications, and the operation of this module is described in more detail below.

3.3.2.1 A/D Converter

The eight A/D converter inputs are named AN0–AN7 and are shared with PORTA and PORTE digital inputs as shown in Figure 3.19. There is only one A/D converter and the analog inputs are multiplexed where only one analog input data is converted to digital at any time. Analog

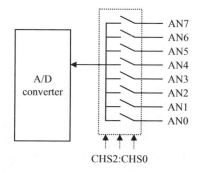

CHS2:CHS0

Figure 3.19 A/D converter block diagram

inputs can directly be applied to these inputs and the A/D converter generates 10-bit digital signals. The A/D module has four registers:

- A/D result high register (ADRESH);
- A/D result low register (ADRESL);
- A/D control register0 (ADCON0);
- A/D control register1 (ADCON1).

The bit definitions of the ADCON0 register are shown in Figure 3.20. This register controls the operation of the A/D converter. The conversion frequency, A/D channels, the A/D status and the conversion command are set by this register.

7	6	5	4	3	2	1	0
ADCS1	ADCS0	CHS2	CHS1	CHS0	GO/DONE	–	ADON

Bit 7-6: ADSC1:ADSC0 A/D converter clock selection
    00: $f_{osc}/2$
    01: $f_{osc}/8$
    10: $f_{osc}/32$
    11: f_{rc}

Bit 5-3: CHS2:CHS0 analog channel select bits
    000: Select channel 0 (AN0)
    001: Select channel 1 (AN1)
    010: Select channel 2 (AN2)
    011: Select channel 3 (AN3)
    100: Select channel 4 (AN4)
    101: Select channel 5 (AN5)
    110: Select channel 6 (AN6)
    111: Select channel 7 (AN7)

Bit 2: GO/DONE A/D conversion status bit
    1: A/D conversion in progress (setting it starts the A/D conversion)
    0: A/D converter not in progress (cleared by hardware when conversion is complete)

Bit 1: Not used

Bit 0: A/D on bit
    1: A/D module is operating
    0: A/D module is shut

Figure 3.20 ADCON0 bit definitions

7	6	5	4	3	2	1	0
ADFM	–	–	–	PCFG3	PCFG2	PCFG1	PCFG0

Bit 7: ADFM A/D result format

    1: Right-justified. Six most significant bits of ADRESH are cleared to 0

    0: Left-justified. Six least significant bits of ADRESL are cleared to 0

Bit 6-4: Not used

    1: Enable EEPROM write complete interrupt

    0: Disable EEPROM write complete interrupt

Bit 3-0: PCFG3-PCFG0 A/D port pin configuration

Figure 3.21   ADCON1 bit definitions

The ADCON1 register configures the functions of the A/D input pins and is used to select the A/D converter reference voltage. The bit definitions of ADCON1 are shown in Figure 3.21. Bit 7 of ADCON1 is called the ADFM bit and controls the format of the converted data. When set to 1, the 10-bit result is right-justified and the six most significant bits of ADRESH are read as 0. When ADFM is cleared to 0, the 10-bit result is left-justified with the six least significant bits of ADRESL read as 0. Bits 0–3 of ADCON1 are used to configure the A/D converter input pins as shown in Figure 3.22.

Note that Vref+ and Vref− in Figure 3.22 are the A/D converter positive and negative reference voltages, respectively. The programmer has the choice of using an external reference voltage, but in most applications Vref+ is programmed to be equal to Vdd (the supply voltage) and Vref− is programmed to be equal to Vss (the supply ground).

The A/D conversion operation must be started by setting the GO/DONE bit of register ADCON0. The end of conversion can be detected in one of two ways. The easiest method is to poll the GO/DONE bit continuously until this bit is cleared. The result is then available in register pair ADRESH:ADRESL. The second method is to program the device to generate interrupts when a conversion is complete.

PCFG3: PCFG0	AN7	AN6	AN5	AN4	AN3	AN2	AN1	AN0	Vref+	Vref−
0000	A	A	A	A	A	A	A	A	VDD	VSS
0001	A	A	A	A	Vref+	A	A	A	RA3	VSS
0010	D	D	D	A	A	A	A	A	VDD	VSS
0011	D	D	D	A	Vref+	A	A	A	RA3	VSS
0100	D	D	D	D	A	D	A	A	VDD	VSS
0101	D	D	D	D	Vref+	D	A	A	RA3	VSS
0110	D	D	D	D	D	D	D	D	VDD	VDD
0111	D	D	D	D	D	D	D	D	VDD	VSS
1000	A	A	A	A	Vref+	Vref−	A	A	RA3	RA2
1001	D	D	A	A	A	A	A	A	VDD	VSS
1010	D	D	A	A	Vref+	A	A	A	RA3	VSS
1011	D	D	A	A	Vref+	Vref−	A	A	RA3	RA2
1100	D	D	D	A	Vref+	Vref−	A	A	RA3	RA2
1101	D	D	D	D	Vref+	Vref−	A	A	RA3	RA2
1110	D	D	D	D	D	D	D	A	VDD	VSS
1111	D	D	D	D	Vref+	Vref−	D	A	RA3	RA2

A = analog input D = digital input

Figure 3.22   A/D converter input pin configuration

The steps required for doing an A/D conversion are listed below:

- Configure the A/D converter module.
  - Configure analog pins and reference voltage.
  - Select A/D input channel.
  - Select A/D conversion clock.
  - Turn on A/D module.
- Configure A/D interrupts (if required).
- Start A/D conversion.
- Set GO/DONE bit of ADCON0 to 1.
- Wait for conversion to complete.
- Poll the GO/DONE bit until it is clear or wait for an A/D interrupt.
- Read A/D converter result from ADRESH and ADRESL.
- Repeat the above as required.

3.4 EXERCISES

1. Explain the basic features common to all PIC microcontrollers.

2. Explain how a PIC microcontroller can be reset using an external push-button switch.

3. Draw the circuit diagram of a minimum PIC system using a 4 MHz crystal.

4. Draw the circuit diagram of a minimum PIC system using a resistor and a capacitor to generate a clock frequency of about 2 MHz.

5. Explain the differences between a crystal and a resonator. Which one would you use if precision timing is required?

6. What is a register file map? Where can it be used?

7. A PIC16F84 microcontroller is to be used in a system. Port A pins RA0 and RA2 are required to be inputs and all port B pins outputs. Determine the values to be loaded into the port A and port B direction registers.

8. Explain the differences between the PIC16F84 and the PIC16F877 microcontrollers.

9. How many banks are there in the PIC16F877? Why are there more banks than in the PIC16F84?

10. Explain where an A/D converter can be used. How many A/D converter channels are there on the PIC16F877? What are the resolutions of these channels?

11. Explain the steps necessary to read data from an A/D channel on the PIC16F877 micro-controller.

12. You are required to generate timer interrupts at $120\,\mu s$ intervals using a 4 MHz crystal as the timing device. Determine the value to be loaded into timer register TMR0. What will be the pre-scaler value?

13. Describe the steps necessary to operate the A/D converter to read analog data from channel AN2. What is the format of this data? How can you right-justify the data?

FURTHER READING

[Anon, 2005a]   Microchip Data on CDROM. Microchip Technology Inc., www.microchip.com.

[Anon, 2005b]   Microchip Databook. Microchip Technology Inc., www.microchip.com.

[Anon, 2005c]   PICC Lite User's Guide. Hi-Tech Software, www.htsoft.com.

[Ibrahim, 2002]   Ibrahim, D. Microcontroller Based Temperature Monitoring and Control. Newnes, London, 2002.

[Iovine, 2000]   Iovine, J. PIC Microcontroller Project Book. McGraw-Hill, New York, 2000.

[James, 1977]   James, M.R. Microcontroller Cookbook PIC and 8051. Newnes, London, 1977.

[Kalani, 1989]   Kalani, G. Microprocessor Based Distributed Control Systems. Prentice Hall, Englewood Cliffs, NJ, 1989.

4
Programming PIC Microcontrollers in C

Microcontrollers have traditionally been programmed using the assembly language. This language consists of various mnemonics which describe the instructions of the target microcontroller. An assembly language is unique to a microcontroller and cannot be used for any other type of microcontroller. Although the assembly language is very fast, it has some major disadvantages. Perhaps the most important of these is that the assembly language can become very complex and difficult to maintain. It is usually a very time-consuming task to develop large projects using the assembly language. Program debugging and testing are also considerably more complex, requiring more effort and more time.

Microcontrollers can also be programmed using the high-level languages. For example, it is possible to use BASIC, PASCAL, FORTRAN and C compilers to program the PIC family of microcontrollers. Most of these compilers generate native machine code which can be directly loaded into the program memory of the target microcontroller.

In this chapter we shall look at the principles of programming PIC microcontrollers using the C language. C is one of the most popular programming languages used today and there are several C compilers available for the PIC microcontrollers. We shall look at how to program the PIC microcontrollers using one of the popular C compilers known as the *PICC Lite* C compiler, developed by the Hi-Tech Software.

PICC Lite is an efficient compiler which can be used to program the PIC16F84, PIC16F877, PIC16F627 and PIC16F629 family of microcontrollers. This compiler is distributed free of charge and is a subset of the PICC programming language, manufactured by Hi-Tech Software. PICC Lite is equipped with an integral editor with syntax highlighting, which makes program development relatively easy. A large number of library functions are provided which can easily be used by the programmer. Some of the reasons for choosing the PICC Lite compiler are:

- support for floating point arithmetic;
- availability of a large number of mathematical functions;
- direct support for LCD displays;
- ease of availability.

As a programming language, C is similar to PASCAL and FORTRAN where values are stored in variables and programs are structured by operating on variables and by defining and calling functions. Program flow is controlled using *if* statements, *while* statements and loops. Inputs can be read from switches, keyboards and sensors, and outputs can be sent to LEDs, LCDs, screens, sound devices, motors and so on. Related data can be stored in arrays or structures.

The program development cycle using C is relatively straightforward. High-level user programs are normally developed on a PC using the integral editor. The code is then compiled and if there are no errors the object code is downloaded into the program memory of the target PIC microcontroller. Depending on the type of microcontroller used, either a flash memory programmer device or an EPROM programmer device is used to load the program memory of the target microcontroller. This cycle is repeated until the developed program operates as required.

4.1   PICC LITE VARIABLE TYPES

The PICC Lite compiler supports the following basic data types:

- bit
- unsigned char
- signed char
- unsigned int
- signed int
- long
- unsigned long
- float
- double

4.1.1   Bit

A *bit* can store a Boolean variable (a 0 or a 1). *Bit* variables are usually used as flags in programs. In the following example, the variable *answer* can only take the values 0 or 1:

```
bit answer;
```

4.1.2   Unsigned Char

An *unsigned char* is 8 bits wide and can store a single ASCII character or an integer number in the range 0 to 255 (bit pattern '11111111'). In the following example, variables *count* and *initial* are declared as *unsigned char* and *count* is assigned a decimal value 120, and *initial* is assigned the ASCII character 'T':

```
unsigned char count, initial;
count = 120;
initial = 'T';
```

4.1.3   Signed Char

A *signed char* (or simply *char*) is 8 bits wide and is used to store signed decimal numbers in the range −128 (bit pattern '10000000') to +127 (bit pattern '01111111'). In the following example, the variable *first* is assigned the decimal value −180, and the variable *count* is assigned a decimal value 25:

```
signed char first, count;
first = -180;
count = 25;
```

4.1.4   Unsigned Int

An *unsigned int* is 16 bits wide and can be used to store decimal numbers in the range 0 to +65 535 (bit pattern '1111111111111111'). In the following example, the variable *result* is assigned the decimal value 28 512:

```
unsigned int result;
result = 28512;
```

4.1.5   Signed Int

A *signed int* (or simply *int*) is 16 bits wide and it is used to store signed numbers in the range −32 768 (bit pattern '1000000000000000') to +32 767 (bit pattern '0111111111111111'). In the following example, the variable *count* is declared as a *signed int* and the negative value − 25 000 is assigned to it:

```
signed int count;
count = -25000;
```

4.1.6   Long

A *long* data type is 32 bits wide and is used to store large signed integer numbers. The range of numbers that can be stored in a *long* are −2 147 483 648 to +2 147 483 647. In the following example, the variable *sum* stores the integer value 45 000:

```
long sum;
sum = 45000;
```

4.1.7   Unsigned Long

An *unsigned long* data type is 32 bits wide and is used to store large unsigned integer numbers. Numbers in the range 0 to 4 294 967 295 can be stored in an *unsigned long* data type. In the following example, the large number 3 200 000 is stored in the variable *cnt*:

```
unsigned long cnt;
cnt = 3200000;
```

4.1.8 Float

A *float* data type is 24 bits wide and is used to store noninteger fractional numbers (i.e. floating-point real numbers). Variables of this type are implemented using the IEEE 754 truncated 24-bit format. In this format a number consists of:

- a 1-bit sign bit;

- an 8-bit exponent which is stored as excess 127 (i.e. an exponent of 0 is stored as 127);

- a 15-bit mantissa. An implied bit to the left of the radix point is assumed, which is always 1, unless the number is zero itself, when the implied bit is 0.

  The value of a number in float representation is given by:

  $$(-1)^{\text{sign}} \times 2^{(\text{exponent}-127)} \times 1.\text{mantissa}$$

  For example, the real number 48.03125 is represented by the following bit pattern:

  `0 10000100 100 0000 0010 0000`

  Here the exponent equals 132, and the mantissa is

  `1 + 2⁻¹ + 2⁻¹⁰ = 1.500 976 562 5.`

  $1 + 2^{-1} + 2^{-10} = 1.500\ 976\ 562\ 5.$

  Multiplying the mantissa by 32 gives the required number 48.03125.
  In the following example, the variable *temp* is loaded with the fractional number 2.35:

```
float temp;
temp = 2.35;
```

4.1.9 Double

A *double* data type is 24 bits or 32 bits wide (selected during the compilation) and is used to store double-precision floating-point numbers. The truncated IEEE 754 24-bit format is used in 24-bit mode. The 32-bit format is based in the IEEE 754 32-bit standard where the number consists of:

- a 1-bit sign bit;
- an 8-bit exponent stored as excess 127;
- a 23-bit mantissa.

In the following example, the variable *temp* stores the number 12.34567 as a double-precision floating-point number:

```
double temp;
temp = 12.34567;
```

4.2 VARIABLES

In C, a variable must be declared before it can be used in a program. Variables are usually declared at the beginning of any block of code. Every variable has a name and a value – the name identifies the variable and the value stores data. There is a rule governing what a variable name can be. Every variable name in C must start with a letter, and the rest of the name can consist of letters, numbers and underscore characters. In C, lower-case and upper-case variable names are different. Also, keywords such as *if*, *while* and *switch* cannot be used as variable names.

Examples of valid variable names are:

```
Xx              sum  result    total  outfile  infile  x1
Xz  Total_result  sum1 highest_no  no_1   cnt       COUNT
```

A variable declaration begins with the data type, followed by the name of one or more variables, and is terminated with the semicolon character. For example,

```
int sum, low, high;
```

Values can be stored in variables after the declaration of the variable types:

```
int low, high, sum;
unsigned char name;
low = 0;
high = 200;
sum = 10;
name = 'R';
```

Variables can also be initialized when they are declared. For example, the above statements can also be written as:

```
int low = 0;
int high = 200;
int sum = 10;
unsigned char name = 'R';
```

4.3   COMMENTS IN PROGRAMS

Comments can be used in programs for clarification purposes. The comment lines are ignored by the compiler. There are two ways of including comments in a program. One way is to use double slashes ('//') before a comment line. Any characters after the double slashes ('//') are ignored by the compiler. For example,

```
j = 0;                          // clear variable j
sum = sum + 1;                  // increment variable sum
//
// The following code clears variables k and m
//
k = 0;
m = 0;
```

Comment lines starting with the '//' characters can be used anywhere in a program.

Another way of creating comment lines in a program is to start the comments with the characters '/*', insert the comments, and then terminate the comments with the '*/' characters.

This method has the advantage that the comments can extend to several lines. An example is given below:

```
/* This program adds two integer numbers
 x and y. The result is stored in variable z.
 z is then incremented by 1
*/
z = x + y;
z = z + 1;
```

The two methods of creating comments can be mixed in a program. For example,

```
/* This program multiplies two integer numbers
   x and y and stores the result in variable z
*/
z = x*y;                             // Multiply the numbers
```

4.4 STORING VARIABLES IN THE PROGRAM MEMORY

In PICC Lite, variables are normally stored in the RAM memory of the target microcontroller since the value of a variable is expected to change during the running of a program. There are many variables whose values do not change during the lifetime of a program, and these variables are known as *constants*. It is possible to store the constants in the flash (or EPROM) program memory of the target microcontroller. The size of the RAM memory is very limited in many microcontrollers and storing some of the variables in the program memory releases valuable RAM memory locations. A variable is stored in the program memory if it is preceded by the keyword const. In the following example, the variable *sum* is stored in the RAM memory, but the variable *pi* is stored in the program memory of the microcontroller:

```
float sum;
const float pi;
```

Any data type can be stored as a constant in the program memory.

4.5 STATIC VARIABLES

Static variables are usually used in functions. A static variable can only be accessed from the function in which it was declared. The value of a static variable is not destroyed on exit from the function, instead its value is preserved and becomes available again when the function is next called. Static variables can be initialized like other normal variables, but the initialization is done once only when the program starts. Static variables are declared by preceding them with the keyword *static*. For example,

```
static int sum;
```

4.6 VOLATILE VARIABLES

A variable should be declared *volatile* whenever its value can be changed by something beyond the control of the program in which it appears, such as an interrupt service routine. The *volatile* qualifier tells the compiler that the value of the variable may change at any time without any action being taken by the code the compiler finds nearby. All I/O based variables and variables shared by the main code and interrupt service routines should be declared volatile. In the following example, the variable *count* is declared volatile:

```
volatile int count;
```

4.7 PERSISTENT VARIABLES

Normally, the initial values of variables are cleared to zero when the program starts up (i.e. whenever reset is applied to the microcontroller). The *persistent* qualifier prevents a variable from being cleared at start-up. In the following example, the variable *max* is declared persistent and as a result its value is not cleared after a reset:

```
persistent int max
```

4.8 ABSOLUTE ADDRESS VARIABLES

It is possible to declare a variable such that it holds the address of an absolute location in the microcontroller memory. This is done by typing the character '@' after the name of the variable. For example,

```
unsigned char Portbit @ 0×06
```

In this code, the variable *Portbit* is assigned the absolute hexadecimal address 0×06. It is important to realize that the compiler does not create a location for the variable *Portbit*, it simply assigns the variable to the specified absolute address.

The *bit* data type and absolute address variables can be used together to access individual bits of a register. For example, the PORTA register is at absolute address 5. We can declare a variable called *PA1* to access bit 1 of PORTA as follows:

```
unsigned char PORTA @ 0×05;
bit PA1 @ (unsigned)&PORTA*8+1;
```

4.9 BANK1 QUALIFIER

The *bank1* type qualifier is used to store static variables in RAM bank 1 of the microcontroller. By default, variables are stored in bank 0 of the RAM. In the following example, the static integer variable temp is stored in bank 1:

```
static bank1 int temp;
```

4.10 ARRAYS

An array is a collection of variables of the same type. For example, an integer array called *results* and consisting of 10 elements is declared as follows:

```
int results[10];
```

As shown in Figure 4.1, each array element has an index and the indices starts at 0. In this example, the first element has index 0 and the last element has index 9. An element of the array is accessed by specifying the index of the required element in a square bracket. For example, the first element of the array is results[0], the second element is results[1], and the last element is results[9]. Any element of an array can be assigned a value which is same type as the type of the array. For example, the integer value 25 can be assigned to the second element of the above array as:

```
results[1] = 25;
```

Initial values can be assigned to all elements of an array by separating the values by commas and enclosing them in a curly bracket. The size of the array should not be specified in such declarations. In the following example, odd numbers are assigned to the elements of array *results* and the size of this array is set automatically to 10:

```
int results[] = {1, 3, 5, 7, 9, 11, 13, 15, 17, 19};
```

The elements of an array can be stored in either the RAM memory or the program memory of a PIC microcontroller. If the array elements never change in a program then they should be stored in the program memory so that the valuable RAM locations can be used for other purposes. Array elements are stored in the program memory if the array name is preceded by the keyword *const*. In the following example, it is assumed that the array elements never change in a program and thus they are stored in the program memory of the microcontroller:

```
const int results[] = {1, 3, 5, 7, 9, 11, 13, 15, 17, 19};
```

Arrays can also store characters. In the following example, the characters COMPUTER is stored in a character array called *device*:

```
unsigned char device[] = {'C', 'O', 'M', 'P', 'U', 'T', 'E', 'R'};
```

| results[0] |
| results[1] |
| results[2] |
| results[3] |
| results[4] |
| results[5] |
| results[6] |
| results[7] |
| results[8] |
| results[9] |

Figure 4.1   An array with 10 elements

In this example, the size of array *device* is set to 8 and the elements of this array are:

```
device[0] = 'C'
device[1] = 'O'
device[2] = 'M'
device[3] = 'P'
device[4] = 'U'
device[5] = 'T'
device[6] = 'E'
device[7] = 'R'
```

Another way of storing characters in an array is as *string* variables. Strings are a collection of characters stored in an array and terminated with the null character (ASCII 0). For example, the characters COMPUTER can be stored as a string in array *device* as:

```
unsigned char device[] = 'COMPUTER';
```

In this example, the array size is set to 9 and the last element of the array is the null character, i.e. the elements of array *device* are:

```
device[0] = 'C'
device[1] = 'O'
device[2] = 'M'
device[3] = 'P'
device[4] = 'U'
device[5] = 'T'
device[6] = 'E'
device[7] = 'R'
device[8] = 0
```

Arrays can have multiple dimensions. For example, a two-dimensional array is declared by specifying the size of its rows followed by the size of its columns. In the example below, a two-dimensional integer array having three rows and five columns is declared. The array is given the name *temp*:

```
int temp[3][5];
```

This array has a total of 15 elements as shown in Figure 4.2. The first element (top left-hand) is indexed as temp[0][0] and the last element (bottom right) is indexed as temp[2][4]. Values can be loaded into the array elements by specifying the row and the column index of the location to be loaded. For example, integer number 50 is loaded into the first row and third column of array *temp*:

```
temp[0][2] = 50;
```

temp[0][0]	temp[0][1]	temp[0][2]	temp[0][3]	temp[0][4]
temp[1][0]	temp[1][1]	temp[1][2]	temp[1][3]	temp[1][4]
temp[2][0]	temp[2][1]	temp[2][2]	temp[2][3]	temp[2][4]

Figure 4.2   A two dimensional array

4.11 ASCII CONSTANTS

The PICC Lite compiler supports a number of ASCII constants which can be used in programs instead of their numeric values. The following are constants the frequently used:

```
\n   newline character
\t   tab character
\\   backslash character
\'   single quote character
\0   null character
```

Character constants are used by enclosing them in single quotes. For example, the variable *temp* can be assigned the null character by writing:

```
temp = '\0';
```

4.12 ARITHMETIC AND LOGIC OPERATORS

The PICC Lite compiler supports a large number of arithmetic and logical operators. The most commonly used operators are summarized below:

```
()[]   parenthesis
!      logical NOT
~      bit complement
+ - * /     arithmetic add, subtract, multiply, divide
%      arithmetic modulo (remainder from integer division)
++     increment by 1
--     decrement by 1
&      address of
<<     shift left
>>     shift right
≥      greater than or equal
>      greater than
≤      less than or equal
<      less than
sizeof      size of a variable (number of bytes)
==     logical equals
!=     logical not equals
||     logical OR
&&     logical AND
+=     add and assign
-=     subtract and assign
/=     divide and assign
|=     logical OR and assign
^=     logical NOT and assign
&=     logical AND and assign
>>=    shift right and assign
<<=    shift left and assign
```

Some of the operators are unique to the C language and need further clarification. The pre-increment and post-increment operators and their equivalents are:

```
++i;                              i = i + 1;
i++;                              i = i + 1;
--i;                              i = i - 1;
i--;                              i = i - 1;
```

It is important to realize that if the operator occurs after a variable name in an expression, the value of the variable is used in the expression, and then the variable is changed afterwards. Similarly, if the operator occurs before the variable name, the value of the variable is changed before evaluating the expression and then the expression is evaluated:

```
        x = a*b++;        is equivalent to    x = a*b;    b = b + 1;
```

but

```
        x = ++b*a;        is equivalent to    b = b + 1;    x = b*a;
```

Similarly,

```
        x = a*b--;        is equivalent to    x = a*b;    b = b -- 1;
```

but

```
        x = --b*a;        is equivalent to    b = b -- 1;    x = b*a;
```

The following short-hand mathematical operations and their equivalents can be used:

```
i += 5;                 is equivalent to    i = i + 5;
i -= 5;                 is equivalent to    i - i - 5;
i *= 5;                 is equivalent to    i = i*5;
i /= 5;                 is equivalent to    i = i/5;
```

A given number can be shifted left or right using the shift operators. '<<' (operator shifts left a number) and '>>' (operator shifts right a number). Assuming that the value of variable *sum* = 8, the shift operations produce the following results:

```
sum = sum << 1;    sum is shifted left 1 digit, new value of sum = 16
sum = sum << 2;    sum is shifted left 2 digits, new value of sum = 32
sum <<= 2;         sum is shifted left 2 digits, new value of sum = 32


sum = sum >> 1;    sum is shifted right 1 digit, new value of sum = 4
sum = sum >> 2;    sum is shifted right 2 digits, new value of sum = 2
sum >> 2;          sum is shifted right 2 digits, new value of sum = 2
```

Logical complement, AND, and OR operations apply on all bits of a variable. Assuming *a* and *b* are 8-bit unsigned character type variables, and a = 6 (bit pattern 00000110) and b = 2 (bit pattern 00000010), the logical operators produce the following results:

```
x = ~ a; x is assigned complement of a. i.e. x
   = 11111001 i.e. x = 249

x = a & b;   x is assigned logical bitwise AND of a and b.
             00000110
             00000010
             --------
             00000010        i.e. the value of x is 2
```

```
x = a | b;   x is assigned logical bitwise OR of a and b.
             00000110
             00000010
             --------
             00000110      i.e. the value of x is 6
```

Logical connectors are used to combine relational operators. For example, the following statement is true only if x is less than 10 and greater than 5:

```
x < 10 && x > 5
```

Similarly, the following statement is true only if either x or y is greater than 20:

```
x > 20 || y > 20
```

In C, these logical connectors employ a technique known as *lazy* evaluation. This means that the expressions evaluate left to right and the right expression is only evaluated if it is required. For example, *false && false* is always false and if the first expression is false there is no need to evaluate the second one. Similarly, *true || true* is always true and if the first expression is true the second one is not evaluated.

An example is given here to clarify the use of the operators.

Example 4.1

sum and *count* and *total* are two 8-bits-wide *unsigned char* type variables. If sum = 65 and count = 240, find the values of the following expressions:

   (i) total = ~sum;
  (ii) total = sum | count;
 (iii) total = sum & count;
  (iv) count + = 1;
   (v) count | = sum;
  (vi) total = count++;
 (vii) total = ++sum;
(viii) total = sum >> 1;
  (ix) total = ~count;
   (x) sum >>= 1.

Solution

  (i) sum = 0100 0001
     total =~ sum = 1011 1110

 (ii) sum   = 0100 0001
    count = 1111 0000
        ----
    total = 1111 0001

(iii) sum   = 0100 0001
    count = 1111 0000
        ----
    total = 1111 0001

(iv) count = count + 1 = 66

 (v) sum   = 0100 0001
      count = 1111 0000

      ----

      total = 1111 0001

(vi) total = count +1 = 241

(vii) total = sum +1 = 66

(viii) sum = 0100 0001, right shift gives total = 0010 0000

(ix) count = 1111 0000
      total = ~ count = 0000 1111

 (x) sum = 0100 0001, right shift gives sum = 0010 0000

4.13 NUMBER BASES

Different number bases can be used in PICC Lite to describe integer numbers. Table 4.1 lists the valid number bases.

4.14 STRUCTURES

Structures are used to collect related items together. For example, the details of a person can normally be stored as

```
unsigned char name[80];
unsigned char surname[80];
unsigned int age;
unsigned char address[80];
```

We can use a structure to declare the above as follows:

```
struct person
{
  unsigned char name[80];
  unsigned char surname[80];
  unsigned int age;
  unsigned char address[80];
};
```

Table 4.1 Number bases

Number base	Format	Example
Binary	0bnumber	0b10000001
Octal	0number	0774
Decimal	number	218
Hexadecimal	0 × number	0 × 7F

A structure does not store any location in the memory when it is declared. We can now declare variables of type *person* with the following statement:

```
struct person my_details;
```

The elements of a structure can be accessed by writing the structure name, followed by a dot, and the element name. For example, the *age* of structure *my_details* can be assigned number 20 as:

```
my_details.age = 20;
```

The *typedef* statement can also be used to declare a structure. For example, the above structure can be defined as:

```
typedef struct {
   unsigned char name[80];
   unsigned char surname[80];
   unsigned int age;
   unsigned char address[80];
   } person;
```

A new variable of type *person* can now be declared as follows:

```
person my_details;
```

The variable name is my_details, it has members called name, surname, age, and address.

Another example of a structure is a complex number consisting of a real number and an imaginary number:

```
typedef struct {
   double real_part;
   double imaginary_part;
} complex;
```

PICC Lite also supports bit fields in structures. Bit fields are allocated starting with the least significant bit of the word in which they will be stored. The first bit allocated is the least significant bit of the byte, and bit fields are always allocated in 8-bit units. In the following example, the structure *flags* consists of a byte where the first bit is named x, the second bit is named y, and the remaining six bits are named z:

```
struct flags
{
   unsigned char x:1;
   unsigned char y:1;
   unsigned char z:6;
};

stcruct flags temp;
```

In this example, the second bit of temp can be cleared as follows:

```
temp.y = 0;
```

4.15   PROGRAM FLOW CONTROL

The PICC Lite language supports the following flow control commands:

- if–else
- for
- while
- do
- goto
- break
- continue
- switch–case

Examples are given in this section which describes the use of these flow control commands.

4.15.1   If–Else Statement

This statement is used to decide whether or not to execute a single statement or a group of statements depending upon the result of a test. There are several formats for this statement, the simplest one is when there is only one statement:

```
if(condition)statement;
```

The following test decides whether a student has passed an exam with a pass mark of 45 and if so, character 'P' is assigned to the variable *student*:

```
if(result > 45)student = 'P';
```

In the multiple-statement version of the *if* statement, the statements are enclosed in curly brackets as in the following format:

```
if(condition)
{
    statement;
    statement;

    ...
    statement;
    statement;
}

For example,

if(temperature > 20)
{
    flag = 1;
    pressure = 20;
    hot = 1;
}
```

The *if* statement can be used together with the *else* statement when it is required to execute alternative set of statements when a condition is not satisfied. The general format is:

```
if(condition)
{
    statement;
    statement;

    ...
    statement;
    statement;
}
else
{
    statement;
    statement;

    ...
    statement;
    statement;
}
```

In the following example, if *result* is greater than 50 the variable *student* is assigned character 'P' and *count* is incremented by 1. Otherwise (i.e. if *result* is less than or equal to 50) the variable *student* is assigned character 'F' and *count* is decremented by 1:

```
if(result > 50)
{
    student = 'P'
    count++;
}
else
{
    student = 'F'
    count--;
}
```

When using the equals sign as a condition, double equals signs '==' should be used as in the following example:

```
if(total == 100)
    x++;
else
    y++;
```

4.15.2 Switch–Case Statement

This is another form of flow control where statements are executed depending on a multi-way decision. The *switch–case* statement can only be used in certain cases where:

- only one variable is tested and all branches depend on the value of that variable;
- each possible value of the variable can control a single branch.

The general format of the *switch–case* statement is as follows. Here, variable *number* is tested. If *number* is equal to *n1*, statements between *n1* and *n2* are executed. If number is equal to *n2*, statements between *n2* and *n3* are executed, and so on. If *number* is not equal to any of the condition then the statements after the *default* case are executed. Notice that each block of statement is terminated with a *break* statement so that the program jumps out of the *switch–case* block.

```
switch(number)
{
 case n1:
  statement;
  ...
  statement;
  break;
 case n2:
  statement;
  ...
  statement;
  break;
 case n3:
  statement;
  ...
  statement;
  break;
 default:
  statement;
  ...
  statement;
  break;
}
```

In the following example, the variable *no* stores a hexadecimal number between A and F and the *switch–case* statement is used to convert the hexadecimal number to a decimal number in the variable *deci*:

```
switch(no)
{
 case 'A':
  deci = 65;
  break;
 case 'B':
  deci = 66;
  break;
 case 'C':
  deci = 67;
  break;
 case 'D':
  deci = 68;
  break;
 case 'E':
  deci = 69;
  break;
```

```
case 'F':
  deci = 70;
  break;
}
```

4.15.3 For Statement

The *for* statement is used to create loops in programs. The *for* loop works well where the number of iterations of the loop is known before the loop is entered. The general format of the *for* statement is:

```
for(initial; condition; increment)
{
 statement;
 ...
statement;
}
```

The first parameter is the initial condition and the loop is executed with this initial condition being true. The second is a test and the loop is terminated when this test returns a false. The third is a statement which is executed every time the loop body is completed. This is usually an increment of the loop counter.

An example is given below where the statements inside the loop are executed 10 times. The initial value of variable *i* is zero, and this variable is incremented by one every time the body of the loop is executed. The loop is terminated when *i* becomes 10 (i.e. the loop is executed 10 times):

```
for(i = 0; i < 10; I++)
{
  sum++;
  total = total + sum;
}
```

The above code can also be written as follows, where the initial value of *i* is 1:

```
for(i = 1; i <= 10; i++)
{
  sum++;
  total = total + sum;
}
```

If there is only one statement to be executed, the for loop can be written as in the following example:

```
for(i = 0; i < 10; i++)count++;
```

It is possible to declare nested *for* loops where one loop can be inside another loop. An example is given below where the inner loop is executed 5 times, and the outer loop is executed 10 times:

```
for(i = 0; i < 10; i++)
{
  cnt++;
```

```
for(j = 0; j < 5; j++)
{
  sum++;
}
}
```

4.15.4 While Statement

The *while* loop repeats a statement until the condition at the beginning of the statement becomes false. The general format of this statement is:

```
while(condition)statement;
```

or

```
while(condition)
{
  statement;
  statement;
  ...
  statement;
}
```

In the following example, the loop is executed 10 times:

```
i = 0;
while(i < 10)
{
  cnt;
  total = total + cnt;
  i++;
}
```

Notice that the condition at the beginning of the loop should become true for the loop to terminate, otherwise we get an infinite loop as shown in the following example:

```
  i = 0;
while(i < 10)
{
  cnt;
  total = total + cnt;
}
```

Here the variable *i* is always less than 10 and the loop never terminates.

4.15.5 Do Statement

This is another form of the *while* statement where the condition to terminate the loop is tested at the end of the loop and, as a result, the loop is executed at least once. The condition to terminate the loop should be satisfied inside the loop, otherwise we get an infinite loop. The general format of this statement is:

```
do
{
  statement;
  statement;
  ...
  statement;
} while(condition);
```

An example is given below where the loop is executed five times:

```
j = 0;
do
{
  cnt++;
  j++;
} while(j < 5);
```

4.15.6 Break Statement

We have seen the use of this statement in *switch–case* blocks to terminate the block. Another use of the *break* statement is to terminate a loop before a condition is met. An example is given below where the loop is terminated when the variable *j* becomes 10:

```
while(i < 100)
{
  total++;
  sum = sum + total;
  if(j = 10)break;
}
```

4.15.7 Continue Statement

The *continue* statement is similar to the *break* statement but is used less frequently. The *continue* statement causes a jump to the loop control statement. In a *while* loop, control jumps to the condition statement, and in a *for* loop, control jumps to the beginning of the loop.

4.16 FUNCTIONS IN C

Almost all programming languages support functions or some similar concepts. Some languages call them subroutines, some call them procedures. Some languages distinguish between functions which return variables and those which do not.

In almost all programming languages functions are of two kinds: user functions and built-in functions. User functions are developed by programmers, while built-in functions are usually general purpose routines provided with the compiler. Functions are independent program codes and are usually used to return values to the main calling programs.

In this section we shall look at both types of functions.

4.16.1   *User Functions*

These functions are developed by the programmer. Every function has a name and optional arguments, and a pair of brackets must be used after the function name in order to declare the arguments. The function performs the required operation and can return values to the main calling program if required. Not all functions return values. Functions whose names start with the keyword *void* do not return any values, as shown in the following example:

```
void led_on()
{
   led = 1;
}
```

The return value of a function is included inside a return statement as shown below. This function is named *sum*, has two integer arguments named *a* and *b*, and the function returns an integer:

```
int sum(int a, int b)
{
   int z;

   z = a + b;
   return(z);
}
```

A function is called in the main program by specifying the name of the function and assigning it to a variable. In the following example, the variable *w* in the main program is assigned the value 7:

```
w = sum(3, 4)
```

It is important to realize that the variables used inside a function are local to that function and do not have any relationship to any variables used outside the function with the same names. An example is given below.

Example 4.2

Develop a function called *area* to calculate the area of a triangle whose base and height are known. Show how this function can be used in a main program to calculate the area of a triangle whose base is 20.2 cm and whose height is 15.5 cm.

Solution

The area of the triangle is given by *base* × *height/2*. The required function is as follows:

```
float area(float base, float height)
{
   float a;
   a = (base * height)/2;
   return(a);
}
```

The function can be used in a main program as follows:

```
my_area = area(20.2,15.5);
```

4.16.2 Built-in Functions

The PICC Lite compiler provides a large number of built-in functions which are available to the programmer. A list of all the available functions can be obtained from the *PICC Lite User's Guide*. Some examples are given below.

4.16.2.1 abs

This function calculates the absolute value of a given number. For example, the absolute value of the variable *total* can be obtained as follows:

```
int sum, total;
sum = abs(total);
```

4.16.2.2 cos

This function returns the trigonometric cosine of an angle. The angle must be in radians. For example, the cosine of 45° can be calculated as follows:

```
float pi, rad, k;
pi = 3.14159;              // value of pi
rad = 45.0 * pi/180;       // convert to radians
k = cos(rad);              // calculate the cosine
```

4.16.2.3 sqrt

The sqrt function returns the square root of a given number. In the following example, the square root of number 14.5 is stored in the variable *no*:

```
double no, k;
no = 14.5;
k = sqrt(no);
```

4.16.2.4 isupper

This function checks whether or not a given character is upper case between A and Z. If it is, a 1 is returned, otherwise a 0 is returned. An example is given below:

```
if(isupper(c))
   b = 10;
else
   b = 2;
```

4.16.2.5 isalnum

This function checks whether or not a given character is an alphanumeric character between 0 and 9, a and z, or A and Z. An example is given below:

```
if(isalnum(c))
  sum++;
else
  sum--;
```

4.16.2.6 strlen

The strlen function returns the length of a string. In the following example, number 3 is returned:

```
k = strlen('tea');
```

4.16.2.7 strcpy

This function copies a string into a character array. In the following example, string 'Computer' is copied into character array buffer:

```
char buffer[80];
strcpy(buffer, 'Computer');
```

4.17 POINTERS IN C

Pointer are widely used in C programs. A pointer is a variable which stores the address of another variable. A pointer is declared by preceding it with the '*' character. For example, a character pointer called *p* is declared as follows:

```
char *p;
```

Similarly, an integer pointer called *pnt* is declared by writing:

```
int *pnt;
```

In the above example, although *pnt* is declared as a pointer, currently it is not assigned any value. The address of a variable is obtained by preceding the variable name with the '&' character. For example, the address of the integer variable *sum* is obtained as follows:

```
pnt = &sum;
```

We can also make the above declaration by writing:

```
int *pnt = &sum;
```

Now, the pointer *pnt* holds the address of the variable *sum*. We can access the value of a variable whose address is known by preceding its pointer with the '*' character. Thus, in the above example, we can set the value of the variable *sum* to 10 as follows:

```
*pnt = 10;
```

In C whenever variables are passed as arguments to a function, their values are copied to the corresponding function parameters, and the variables themselves are not changed in the calling environment. This is referred to as the *call-by-value* mechanism. Most other languages provide call-by-reference mechanisms so that the values of variables can be changed in functions. In C this is done using pointers where the addresses of the variables are passed to a function and not their values. An example function is given below which swaps the values of two of its arguments:

```
void swap(int *p, int *q)
{
    int temp;
    temp = *p;            // store value of p in temp
    *p = *q;              // store value of q in p
    *q = temp;            // store temp in q
}
```

The main program then calls the function as follows:

```
swap(&m, &n);
```

Notice that in the main program the addresses of variables are passed to the function and the function definition uses pointers.

Pointers are frequently used in arrays and in function arguments. An array is actually a pointer to the zeroth element of the array. The array name gives the address of an array. By using pointer arithmetic we can access the elements of array easily. An example is given below:

```
char buffer[10];         // declare a character array
char *p;                 // declare a character pointer
p = buffer;              // p holds the address of buffer[0]
*p = 0;                  // clear first array element (buffer[0] = 0)
p++;                     // increment pointer (point to buffer[1])
*p = 10;                 // set buffer[1] = 10;
```

Thus, if p = buffer, then $*(p + 1)$ points to buffer[1], $*(p + 2)$ points to buffer[2], and in general $*(p + n)$ points to array element buffer[n].

Since an array is like a pointer, we can pass an array to a function, and modify elements of that array without having to worry about referencing the array. For example, an integer array called *sum* can be passed to a function as follows:

```
int sum[];
```

or

```
int *sum;
```

Either of these definitions is independent of the size of the array being passed, and the array size is in general not known by the function. In a function definition, a formal parameter that is defined as an array is actually a pointer. When an array is passed as an argument to a function, the base address is passed and the array elements themselves are not copied. An array bracket notation is used to declare pointers as parameters. An example is given below where it is assumed that array *a* has 10 elements:

```
int sum(int a[])
{
```

```
   int total = 0;
   int i;

   for(i = 0; i < 10; i++)total = total + a[i];
   return(total);
}
```

and the function can be called from the main program as:

```
sum(t);
```

where array *t* is an integer array with 10 elements, i.e. int t[10].

In the header of the function, *a* is declared as an array and this is equivalent to declaring it as a pointer, i.e. the above function can also be written as:

```
int sum(int *a)
{
   int total = 0;
   int i;

   for(i = 0; i < 10; i++)total = total + a[i];
   return(total);
}
```

or as

```
int sum(int *a)
{
   int total = 0;
   int i;

   for(i = 0; i < 10; i++)total = total + *(a+i);
   return(total);

}
```

4.18   PRE-PROCESSOR COMMANDS

Lines that begin with a '#' character in column 1 of a C program are called pre-processor commands. The C language relies on the pre-processor to extend its power. Some of the commonly used pre-processor commands are described in this section.

4.18.1   #define

This command is used to replace numbers and expressions with symbols. Some examples are given below:

```
#define        PI          3.14159
#define        MAX         1000
#define        MIN         0
```

When these symbols are used in the program their values are substituted to where the symbols are. For example,

```
if(result > MAX)
```

is changed by the pre-processor into

```
if(result > 1000)
```

#define commands can be complex, such as:

```
#define        MULT(a, b)        2*(a + b)
#define        SECONDS           (60* 60* 24)
```

In the last example, the pre-processor will replace every occurrence of SECONDS by the string (60* 60* 24).

Other examples of *#define* are:

```
#define        SQ(x)             ((x) * (x))
```

where SQ(x) will be replaced with ((x) * (x)). For example, SQ(a + b) will be replaced with ((a + b) * (a + b)).

#define can also be used to define macros. A macro is frequently used to replace function calls by in-line code, which is more efficient. An example is given below:

```
#define        MIN(x, y)         (((x) < (y)) ? (x): (y))
```

which compares x and y and if x is less than y, the result is x, otherwise the result is y. After this definition, an expression such as:

```
k = MIN(u, v);
```

gets expanded by the pre-processor to:

```
k = (((u) < (v)) ? (u): (v));
```

Example 4.3

An integer consists of two bytes. Write a macro to extract the upper and the lower bytes of an integer. Call the upper byte *ubyte* and the lower byte *lbyte*.

Solution

We can shift right eight times to extract the upper byte:

```
#define ubyte(x)            (unsigned char)(x >> 8)
```

The lower byte can be extracted by masking the integer with hexadecimal number 0xFF.

```
#define lbyte(x)            (unsigned char)(x & 0xff)
```

Notice that because the x is an integer, the result is converted into a character by preceding it with '(unsigned char)'. This is called *casting* in C.

Example 4.4

Write a macro to set or clear a bit of a variable.

Solution

If we name the macro to set a bit as *bit_set*,

```
#define      bit_set(var, bitno)      ((var) = (1 ≪ (bitno)))
```

and the macro to clear a bit as *bit_reset*,

```
#define      bit_reset(var, bitno)    ((var) &= ~(1 ≪ (bitno)))
```

we can now set bit 3 of the variable x to 1 with the following statement:

```
bit_set(x, 3);
```

Similarly, bit 6 of the variable q can be cleared to 0 with the following statement:

```
bit_set(q, 6);
```

4.18.2   #include

This command causes the pre-processor to replace the line with a copy of the contents of the named file. For example,

```
#include <myproject.h>
```

or

```
#include 'myproject.h'
```

causes the contents of file myproject.h to be included at the line where the command is issued.

When developing programs with the PICC Lite compiler, the following line must be included at the beginning of all programs:

```
#include <pic.h>
```

The file pic.h includes all the PIC microcontroller register definitions.

4.18.3   #asm and #endasm

These pre-processor commands are used to include assembler instructions in C programs. The *#asm* command identifies the beginning of the assembly code, and *#endasm* identifies the end of the assembly code. For example,

```
...
...
i = 10;
#asm
movlw 10h
#endasm
...
...
```

4.19   ACCESSING THE EEPROM MEMORY

Some PIC microcontrollers (e.g. PIC16F84) have EEPROM memories which can be used to store nonvolatile data. PICC Lite provides instructions for reading and writing to this memory. The EEPROM_WRITE command is used to write a data byte to the EEPROM memory. For example, the following command writes 0x2E to address 2 of the EEPROM memory:

```
EEPROM_WRITE(2,0×2E);
```

Similarly, the EEPROM_READ command is used to read a data byte from the EEPROM memory. For example, the following command reads a byte from address 5 of the EEPROM memory and loads it into the variable k:

```
k = EEPROM_READ(5);
```

4.20   INTERUPTS IN C PROGRAMS

Interrupts are very important in real-time programs. Most PIC microcontrollers offer internal and external interrupt facilities. Internal interrupts are usually timer-generated interrupts, while the external interrupts are triggered by activities on the I/O pins. When an interrupt occurs the processor jumps to the interrupt service routine where the interrupt is serviced. Medium-range microcontrollers (such as the PIC16F84) have only one ISR, which is at address 0x04 of the program memory. If interrupts from multiple sources are expected then the interrupting device can be found by inspecting the INTCON register bits.

Interrupts can be used in PICC Lite programs in the form of special functions. When an interrupt occurs the program jumps to this special function where the interrupt is handled. ISR function must have the name *interrupt* followed by a user-selected function name, and the function must not return any value and must not have any arguments. For example, an ISR function named *my_int* is given below:

```
void interrupt my_int(void)
{
   statement;
   statement;
   ...
   statement;
}
```

Interrupts must be enabled before they can be recognized by the microcontroller. The *ei*() and *di*() instructions enable and disable global interrupts, respectively. Additionally, the interrupt control bit of each interrupt source must be enabled in the appropriate SFR register (e.g. T0IE bit in INTCON must be set to 1 to enable timer interrupts).

The PIC microcontroller only saves the program counter on stack when an interrupt occurs. The PICC Lite compiler determines which registers and objects are used by the interrupt function and saves these appropriately.

4.21 DELAYS IN C PROGRAMS

There are many real-time applications where the microcontroller is too fast and we need to slow it down to see, for example, the flashing of an LED. The PICC Lite compiler offers two types of delay functions: a microsecond delay function, and a millisecond delay function. The file <delay.c> must be included at the beginning of a program before these delay functions can be used. The function *DelayUs(x)* is used to create delays in the range 1 to 255 μs. For example, the following function call generates a 200 μs delay:

```
DelayUs(200);
```

Similarly, function *DelayMs(x)* is used to create delays in the range 1 to 255 ms. The following function call creates a 10 ms delay:

```
DelayMs(10);
```

Longer delays can be created by using the above delay functions in loops. These delay functions are by default based on the assumption that the microcontroller operates with a 4 MHz clock frequency. If the clock rate is different, then the new clock rate must be specified at the beginning of the program. For example, if the clock rate is 2 MHz, then the following lines must be included at the beginning of the program:

```
#undef XTAL_FREQ
#define XTAL_FREQ 2MHZ
```

4.22 STRUCTURE OF A C PROGRAM

The simplest C program consists of three lines:

```
main(void)
{
}
```

main is the starting point of the main program and the statements of the main program are written inside a pair of curly brackets.

A more useful program which calculates the average of three numbers 5, 7 and 9 is given below:

```
main(void)
{
  float average;
  average = (5 + 7 + 9)/3;
}
```

It is always a good practice to include comments in programs to describe the operation of the various parts of the program. Comment lines should also be included at the beginning of a program to describe the aim of the program, the author, filename, date, and any modifications.

Also, the file <pic.h> (in the compiler's include directory) should be included in all PICC Lite programs. An example is given below:

```
//    ***********************************************************
//
//                         AVERAGE PROGRAM
//                         ==================
//
// Author: D. Ibrahim
// Date: February 2005
// File: AVERAGE.C
//
// This program calculates the average of three numbers 3, 5, and 9
//
//    ***********************************************************
#include <pic.h>
//
// Main Program
// ************
//
   main(void)
  {
    float average;
    average = (3 + 5 + 9)/3;
  }
```

When functions are used in programs, they should normally be included before the main program. An example is given below.

Example 4.5

Write a program to calculate the square of a number. The square should be calculated using a function.

Solution

```
//    ***********************************************************
//
//                         SQUARE PROGRAM
//                         ================
//
//   Author: D. Ibrahim
//   Date: February 2005
//   File: SQUARE.C
//
//   This program calculates the square of an integer number
//
//    ***********************************************************
#include <pic.h>
//
// Functions
```

```
// ********
    int square_of_number(int x)
    {
       int k;
       k = x * x;
       return(k);
    }
// Main Program
// ************
//
    main(void)
    {
       int p, z;

       p = 5;
       z = square_of_number(p);
    }
```

4.23   PIC MICROCONTROLLER INPUT–OUTPUT INTERFACE

A microcontroller communicates with the outside world using its input–output ports. These ports can be analog or digital. Analog ports are used to read data from analog devices, e.g. the voltage across a resistor. Digital ports are used to read data from digital devices, e.g. the state of a switch. Microcontroller outputs are usually connected to LEDs, LCDs, buzzers, seven-segment displays and similar devices. The input can be a push-button switch, a keyboard or a similar device.

PIC microcontroller output ports can source and sink 25 mA of current. When an output port is sourcing current, the current flows out of the port pin. Similarly, when an output port is sinking current, the current flows towards the output pin. Devices such as LEDs, LCDs and other small devices which operate with up to 25 mA can be connected to the microcontroller output ports. Devices which require larger currents can be connected to the microcontroller ports by using a transistor switch to increase the current capability of the port pin.

4.23.1   Connecting an LED

Most LEDs operate with about 2V and they draw about 10 mA from the power supply. An LED can be connected to the output port of the PIC microcontroller by using a series current limiting resistor, as shown in Figure 4.3.

Assuming the output voltage V_o of the port pin is +5 V when it is at logic 1, the value of the resistor can be calculated as

$$R = \frac{V_o - 2}{10\,mA} = \frac{5V - 2V}{10\,mA} \approx 330\,\Omega.$$

Example 4.6

An LED is connected to bit 0 of port B (i.e. pin RB0) of a PIC16F84 microcontroller. Write a program to flash the LED at 100 ms intervals. The hardware set-up is shown in Figure 4.4, where the microcontroller is operated with a 4 MHz crystal.

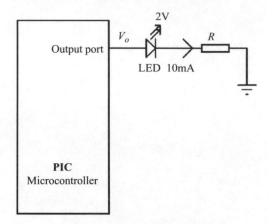

Figure 4.3   Connecting an LED

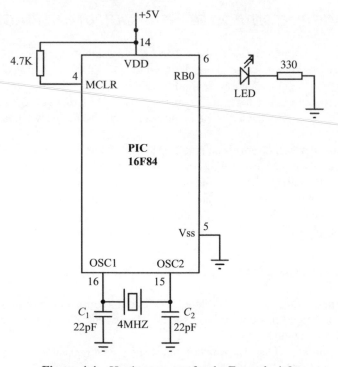

Figure 4.4   Hardware setup for the Example 4.6

Solution

The required program is given in Figure 4.5. At the beginning of the program port B is configured as an output port. The LED is then turned on by sending a logic 1 to the port pin (RB0 = 1). After 100 ms delay the LED is turned off (RB = 0) and this cycle is repeated forever.

```
//********************************************************************
//
// LED FLASHING PROGRAM
// ====================
//
// Author:   D. Ibrahim
// Date:     May, 2005
// File:     FLASH.C
//
// This program flashes an LED conencted to port RB0 of a PIC16F84
// microcontroller.
// The microcontroller is operated with a 4MHz crystal.
//
//********************************************************************
#include <pic.h>
#include <delay.c>
//
// Main Program
// ************
//
main(void)
{
 TRISB = 0;                 /* PORT B is output */

 for(;;)                    /* Do FOREVER */
 {
  RB0 =1;                   /* Turn ON LED */
  DelayMs(100);             /* 100ms delay */
  RB0 = 0;                  /* Turn OFF LED */
  DelayMs(100);             /* 100ms delay */
 }
}
```

Figure 4.5   Program to flash the LED

4.23.2 *Connecting a push-button switch*

A push-button switch is an input to the microcontroller. The simplest way to connect a switch is to connect the switch to one of the port pins of the microcontroller and pull up the port pin to the supply voltage +V using a resistor. The port pin is thus normally at logic 1 level. When the switch is pressed the port pin can be shorted to ground to make the input go to logic 0. An example is given below.

Example 4.7

An LED is connected to bit 0 of port B (i.e. pin RB0) of a PIC16F84 microcontroller. Also, a push-button switch is connected to bit 7 of port B (i.e. pin RB7) using a resistor. Write a program which will turn ON the LED when the switch is pressed. The hardware set-up is shown in Figure 4.6.

Solution

The required program is given in Figure 4.7. At the beginning of the program port pin RB0 is configured as an output and port pin RB7 is configured as an input (bit pattern

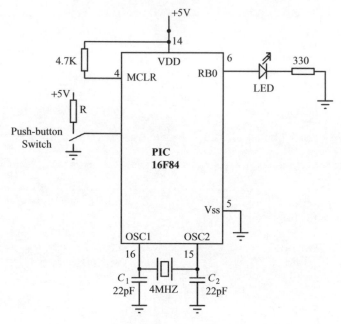

Figure 4.6   Hardware setup for Example 4.7

```
//*********************************************************************
//
// PUSH-BUTTON AND LED PROGRAM
// ===========================
//
// Author:  D. Ibrahim
// Date:     May, 2005
// File:     BUTTON.C
//
// This program turns ON an LED when a push-button switch is pressed.
// The LED is conencted to port pin RB0 and the switch is connected
// RB7.  The microcontroller is operated with a 4MHz crystal.
//
//*********************************************************************
#include <pic.h>
//
// Main Program
// ************
//
main(void)
{
  TRISB = 0x80;            /* RB0 is output, RB7 is input */

  RB0 = 0;                 /* Make sure the LED is OFF when started */
  while(RB7 == 1);         /* Wait until switch is pressed */
  RB0 = 1;                 /* Turn ON LED */
}
```

Figure 4.7   Program for Example 4.7

10000000 = 0 × 80 is sent to TRISB register). The state of the switch is then checked continuously and as soon as the switch is pressed the LED is turned on.

4.23.3 Connecting an LCD

LCD displays are commonly used in microcontroller based systems to display the value of a variable, to prompt the user for data, or to give information to the user. LCDs can either be text based or graphical. Text based LCDs are used in most microcontroller applications. These LCDs are easier to program and their costs are much lower than graphical displays.

One of the most popular LCD displays is based on a controller known as the HD44780. There are several models of LCDs using this controller:

LM016L   2 rows × 16 characters per row
LM017L   2 rows × 20 characters per row
LM018L   2 rows × 40 characters per row
LM044L   4 rows × 20 characters per row

The programming of an LCD is generally a complex task and the programmer needs to know the internal operations of the LCD controller. Fortunately, the PICC language supports the HD44780 type LCDs and any data can easily be displayed on an LCD using simple function calls. The following functions are available:

lcd_init   initialize the LCD
lcd_clear   clear the LCD and home the cursor
lcd_goto   go to the specified cursor position
lcd_write   send a character to the LCD
lcd_puts   send a text string to the LCD

HD44780 type LCDs normally have 14 pins. Table 4.2 shows the pin numbers and the function of each pin. Pin 3 is used to control the contrast of the display. Typically this pin is connected to the supply voltage using a potentiometer, and the contrast is changed by moving the arm of the potentiometer. The RS pin is used to send a control message or a text message to the LCD. When the R/W pin is at logic 0, a command or a text message can be sent to the LCD, and this is the normal operating mode. When R/W is at logic 1, the LCD status can be read. The LCD is enabled when the E pin is at logic 0. Pins D0 to D7 are the data inputs. The LCD can either be used in full 8-bit mode, or in 4-bit half mode where only the upper four data pins are used. In most applications the 4-bit mode is selected since it uses fewer pins and frees the microcontroller input–output pins. The PICC language configures the LCD in 4-bit mode.

In order to use the above LCD functions, an LCD must be connected in a certain way to the microcontroller port pins. The default connection is:

Port pin	LCD pin
RB0	D4
RB1	D5
RB2	D6
RB3	D7
RA2	RS
RA3	E

Table 4.2   CD pin configuration

Pin no.	Name	Function
1	Vss	Ground
2	Vdd	+V supply
3	Vee	Contrast control
4	RS	Select
5	R/W	Read/write
6	E	Enable
7	D0	Data 0
8	D1	Data 1
9	D2	Data 2
10	D3	Data 3
11	D4	Data 4
12	D5	Data 5
13	D6	Data 6
14	D7	Data 7

This connection can be changed by modifying the LCD configuration file <lcd.c> supplied by the PICC compiler.

Example 4.8

An LCD is connected to a PIC16F84 microcontroller as shown in Figure 4.8. Write a program to display the string 'CONTROL' on the LCD.

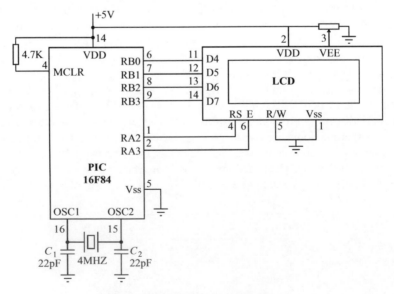

Figure 4.8   Connecting an LCD

Solution

The program listing is shown in Figure 4.9. At the beginning of the program ports A and B are configured as outputs. Then the LCD is initialized and the string 'CONTROL' is sent to the LCD using the command *lcd_puts*.

A more complex microcontroller example using an analog temperature sensor, an A/D converter and an LCD is given below.

Example 4.9

An LM35DZ type analog temperature integrated circuit is connected to analog input AN0 (or RA0) of a PIC16F877 microcontroller. Also, an LCD is connected to the microcontroller as shown in Figure 4.10. Write a program to display the ambient temperature every second on the LCD. The display should show the temperature as 'TEMP = nn', where nn is the ambient temperature. This is an example of a digital thermometer.

```
//*********************************************************************
//
// LCD DISPLAY PROGRAM
// ===================
//
// Author:   D. Ibrahim
// Date:     May, 2005
// File:     LCD.C
//
// This program sends the message CONTROL to an LCD.
// The LCD is connected to a PIC microcontroller as specified by the
// PIC C language compiler. The microcontroller is operated with a
// 4MHz crystal.
//
//*********************************************************************
#include <pic.h>
#include <delay.c>
#include <lcd.c>
//
// Main Program
// ************
//
main(void)
{
  TRISA = 0;                      /* PORT A is output */
  TRISB = 0;                      /* PORT B is output */

  lcd_init();                     /* Initialize the LCD */
  lcd_clear();                    /* Clear the display */
  lcd_puts("CONTROL");            /* Send text CONTROL to the LCD */
}
```

Figure 4.9   Program listing for Example 4.8

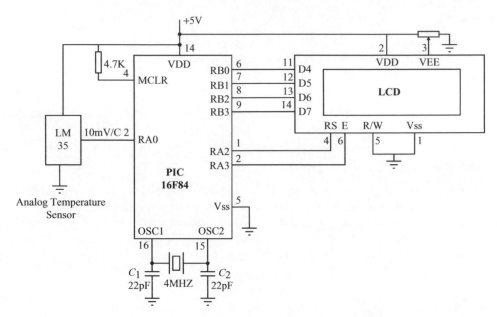

Figure 4.10   Hardware set-up of the digital thermometer

Solution

The LM35DZ is a 3-pin integrated circuit which gives an output voltage which is directly proportional to the temperature. The device can be used to measure temperatures in the range 0–125° C. Two pins of the integrated circuit are connected to the supply and the ground. The third pin is the output V_o, where $V_o = 10 \, \text{mV}/° C$. Thus, for example, at 20 °C the output voltage is 200 mV, at 30°C the output is 300 mV and so on. The sensor is directly connected to analog channel AN0 of the microcontroller. The LCD is connected as described in Example 4.8.

The program listing of the digital thermometer is given in Figure 4.11. At the beginning of the program bit 0 of port A is configured as input and port B is configured as output. Then, ADCON1 register is configured (see Chapter 3) by sending the bit pattern '10001110 = 0 × 8E', so that the RA0 port pin is analog, and the RA2 and RA3 port pins are digital. Also, bit 7 (ADFM) of ADCON1 is set to 1 so that the 8 bits of the converted data will be in the register ADRESL and the upper 2 bits will be in bits 0 and 1 of the register ADRESH. The A/D converter clock is chosen as $f_{osc}/8$ and channel 0 of the A/D converter is selected by configuring the ADCON0 register.

The A/D converter is started by setting bit 2 of the ADCON0 register (sending 0 × 45 to ADCON0). The program then waits (using a *while* loop) for the completion of the conversion. At the end of the conversion bit 2 of ADCON0 is cleared by the microcontroller and this is sensed by the program. The program then exits the *while* loop and reads the lower and upper bytes of the converted 10-bit data, combines the data into a single variable called *temp*, and then converts the data into real value, i.e. millivolts. The actual temperature in degrees Celsius is then obtained by dividing this voltage by 10. The temperature is then converted into a string called *tempc* and is sent to the LCD display using the *lcd_puts* command. The program clears

```
//**********************************************************************
//
// DIGITAL THERMOMETER PROGRAM
// ===========================
//
// Author:  D. Ibrahim
// Date:    May, 2005
// File:    THERMOMETER.C
//
// This program reads the temperature form a LM35DZ type analog
// sensor every second and then displays the temperature on an LCD.
// The microcontroller is operated with a 4MHz crystal.
//
//**********************************************************************
#include <pic.h>
#include <delay.c>
#include <lcd.c>
#include <stdio.h>

//
// Function to wait a second
//
void wait_a_second()
{
 unsigned int j;
 for(j = 0; j < 4; j++)DelayMs(250);
}



//
// Main Program
// ************
//
main(void)
{
 const float lsb = 5000.0/1024.0;
 float mV, temp, templ, temph;
 unsigned int tempc;
 unsigned char disp[] = "TEMP =    ";

 TRISA = 1; /* RA0 is input, others output */
 TRISB = 0; /* PORT B is output */

 ADCON1 = 0x8E; /* RA0=analog, RA2,RA3=digital */
 ADCON0 = 0x41; /* Configure A/D clock and select channel 0 */

 for(;;)
 {
  ADCON0 = 0x45; /* Start A/D conversion */
  while(ADCON0 & 4) != 0); /* Wait for conversion */
  temph = ADRESH; /* Read upper 2 bits */
  templ = ADRESL; /* Read lower 8 bits */
  temp = 256.0*temph + templ; /* Temperature in digital */
  mV = temp *lsb; /* Temperature in mV */
  tempc = mV / 10.0; /* Temperature in Centig. (10mV/C) */
  sprintf(disp+7, "%d",tempc); /* Convert temperature to a string */
  lcd_puts(tempc); /* Display temperature on LCD */
  wait_a_second(); /* Wait a second */
  lcd_clear(); /* Clear display */
 }
}
```

Figure 4.11   Program listing for Example 4.9

the LCD display and repeats continuously after 1 s delay. The delay is created by using a function called *wait_a_second*. The program displays the temperature as $TEMP = nn$ where nn is the ambient temperature.

Notice that the program uses the header files <pic.h>, <delay.c>, <lcd.c>, and <stdio.h>. The file <pic.h> contains the PIC microcontroller definitions. <delay.c> is used to create delays in the program. <lcd.c> is used for the LCD initialization and control functions. Finally, <stdio.h> is used to convert an integer into a string so that it can be displayed on the LCD.

4.24 EXERCISES

1. Describe the differences between an *unsigned char* and a *signed char*.

2. Describe the differences between an *int* and a *float*.

3. Show how integer number 230 can be written in binary, octal and in hexadecimal.

4. Write a function which converts the temperature from °F to °C.

5. What will be the value of the variable z in the following program?

```
int f(x)
{
    int p;
    p = x++;
    return(p);
}

main(void)
{
    int x, z;
    x = 10;
    z = f(x);
}
```

6. The mathematical operation min(x, y) can be represented by the conditional expression

```
(x < y)?x:y
```

   In a similar fashion, describe the mathematical operation min(x, y, z).

7. Write a function that reverses the bit pattern of a byte. For example,

```
00000001     reversing gives 1111110
```

8. Write a function which takes a variable x and shifts it n positions to the left.

9. Write the octal, hexadecimal and binary equivalents of the decimal integers 4, 6, 120, 254.

10. Write a function to multiply two polynomials of degree n.

11. Write a program which calls a function to calculate the sum of the squares of the elements of a matrix which has 10 elements.

12. Explain how delays can be generated in PICC Lite programs. Describe how a 10 s delay can be generated using *do* loops.

13. Explain how data can be written to the EEPROM memory of a PIC microcontroller. Write a program to clear all the first 64 locations of the EEPROM memory.

14. Explain how data can be read from the EEPROM memory of a PIC microcontroller. Write a program to calculate the sum of all numbers stored in the first 20 locations of an EEPROM memory.

15. Eight LEDs are connected to port B of a PIC16F84 microcontroller. Write a program to turn on the odd-numbered LEDs, i.e. LEDs connected to port pins 1, 3, 5 and 7.

16. Eight LEDs are connected to port B of a PIC16F84 microcontroller. Also, a push-button switch is connected to port RA0. Write a program to turn on the LEDs clockwise when the switch is pressed, and anti-clockwise when the switch is released.

17. An LCD is connected to a PIC16F84 microcontroller. Write a program to display the string 'My Computer' on the LCD.

18. A PIC16F84 microcontroller is to be used as an event counter. An LCD is connected to the microcontroller in the standard way, and a switch is connected to bit 0 of port A. The switch is designed to close every time an external event occurs. Write a program which will count the events and display the total on the LCD continuously.

19. A PIC16F877 microcontroller is to be used to monitor the temperature of an oven and give an alarm if the temperature exceeds a pre-specified value. An LM35DZ type analog temperature sensor is to be used to measure the temperature of the oven every second. An LED is connected to bit 0 of port b. (a) Draw the circuit diagram of the system. (b) Write a program that will turn on the LED if the oven temperature exceeds 50°C.

FURTHER READING

[Anon, 2005a]	Microchip Data on CDROM. Microchip Technology Inc., www.microchip.com.
[Anon, 2005b]	PICC Lite User's Guide. Hi-Tech Software, www.htsoft.com.
[Gomaa, 1984]	Gomaa, H. A software design method for real-time systems. *Commun. ACM*, **27**, 1984, pp. 938–949.
[Gomaa, 1986]	Gomaa, H. Software development for real-time systems, *Commun. ACM*, **29**, 1986, pp. 657–668.
[Ibrahim, 2002]	Ibrahim, D. Microcontroller Based Temperature Monitoring and Control. Newnes, London, 2002.
[Iovine, 2000]	Iovine, J. PIC Microcontroller Project Book. McGraw-Hill, New York, 2000.
[James, 1977]	James, M.R. Microcontroller Cookbook PIC and 8051. Newnes, London, 1977.
[Kalani, 1989]	Kalani, G. Microprocessor Based Distributed Control Systems. Prentice Hall, Englewood Cliffs, NJ, 1989.
[Kelley and Pohl, 1984]	Kelley, A.L. and Pohl, I. A Book on C. The Benjamin/Cummings Publishing Co. Inc., Menlo Park, CA, 1984.

5

Microcontroller Project Development

5.1 HARDWARE AND SOFTWARE REQUIREMENTS

The development of a microcontroller based project requires hardware and software products. Hardware requirements generally depend on how complex the project is, but the following hardware products are normally required in almost all types of projects:

- microcontroller programmer;
- microcontroller development board or breadboard with the required components;
- microcontroller chips;
- PC;
- test equipment such as a voltmeter, logic pulser or oscilloscope.

Figure 5.1 shows the basic hardware requirements. A microcontroller programmer is connected to a PC and is used to download the user program to the target microcontroller program memory. For flash type program memories no additional hardware is normally required. The development of systems based on EPROM type program memories requires an EPROM eraser device so that the microcontroller program memory can be erased and reprogrammed.

Small projects incorporating simple LEDs and buzzers can be developed using microcontroller development boards. These boards usually have built-in LEDs, switches, buzzers, etc. so that the user can test programs with simple to moderate complexity. Some development boards also incorporate chip programmer hardware so that the target microcontroller can be programmed on the same board. Complex projects can initially be built and tested on a breadboard. If the project is to be used in commercial or in industrial applications, then a printed circuit board design of the project is created.

A PC is required mainly for two purposes during the development of a microcontroller based project: the user program is developed and compiled on the PC, and the PC is used to transfer the user object code to the device programmer so that the program memory of the target microcontroller can be loaded with the user program.

Depending upon the complexity of the project, several types of test equipment may be required. For simple projects a voltmeter may be sufficient to test the static voltage levels around the circuit. For more complex projects, a logic analyser, logic pulser, frequency counter, or an oscilloscope may be required.

Microcontroller Based Applied Digital Control   D. Ibrahim
© 2006 John Wiley & Sons, Ltd

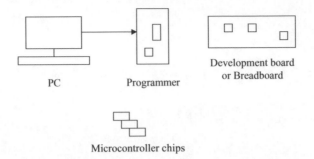

Figure 5.1   Microcontroller project development hardware

In addition to the above hardware, a number of software products will be required during the development of a microcontroller based product. The minimum required software is:

- program development software;
- microcontroller assembler (or compiler if a high-level language is used);
- microcontroller device programmer software.

Program development software, or an editor, is required to write the program code. Most assemblers or compilers provide built-in editors and the user programs can be developed using these editors.

Once a program is written it has to be assembled or compiled if a high-level language is used. The object code is normally produced from the assembler or the compiler if the program contains no errors.

A microcontroller device programmer software is then required to transfer the object code to the program memory of the target microcontroller.

Depending on the complexity of the project, additional software products, such as simulators, debuggers or in-circuit emulators, can be used to test and verify the operation of a program.

Simulator programs are run on a PC and can be used without any project hardware. Simulators are extremely useful in single-stepping and testing the user programs before the program is loaded into the target hardware. Debuggers are similar to simulators, and some debuggers require the code to be loaded into the target microcontroller. The user can insert break-points using debuggers and then test the flow of data and control in a program.

In-circuit emulators can be used in complex projects. Using an emulator, the user can test a program very easily on the target hardware by inserting break-points, and by single-stepping using the target hardware. Although the in-circuit emulators can be very useful, they are usually very expensive.

5.2   PROGRAM DEVELOPMENT TOOLS

Historically, modular programming has been accepted as a good software design concept. Also known as structured programming, a software task is divided into smaller manageable tasks where each task is a self-contained piece of code, also called a module. Modules are then designed using well-known constructs for sequence, selection and iteration. Although a structured approach to programming does not guarantee that a program will be free of errors, it helps to minimize the design errors and makes the final code much more readable and maintainable.

There are many tools available to help the programmer in the development and design of good programs. Some popular tools are: flow charts, structure charts, Unified Modeling Langauge™, Nassi–Schneidermann diagrams, Ferstl diagrams, Hamilton–Zeldin diagrams, and pseudocode. In this section we shall only look at some of the commonly used techniques. Further detailed information can be obtained from most books and papers on computer science.

5.2.1 Flow Charts

Flow charts have been around since the early days of programming. These type of charts are only useful for small applications. One of the disadvantages of flow charts is that the drawing and modifying the diagrams can be very time- consuming. Flow charts also have the disadvantage that they tend to produce unstructured code which is very difficult to maintain. A typical flow chart is shown in Figure 5.2.

5.2.2 Structure Charts

Structure charts, also known as *Jackson structured programming tools*, were developed in the 1970s by Michael Jackson and became a widely used software design tool, especially in Europe.

Structure charts are similar to flow charts but are easier to draw and modify. Structure charts also tend to produce well-structured code which is easy to understand and maintain. The three basic operations of sequence, selection and iteration are shown differently using structure charts.

5.2.2.1 Sequence

Sequence is shown with rectangles drawn next to each other. The sequence of operations is from left to right. An example is given in Figure 5.3 where first the I/O port is initialized, then the LED is turned on, and finally the LED is turned off after a 5 s delay.

5.2.2.2 Selection

Selection is shown by placing a small circle at the top right-hand side of a rectangle. An example is given in Figure 5.4 where if *condition1* is true then process B is performed, and if *condition2* is true process C is performed.

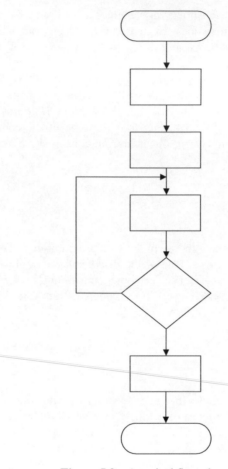

Figure 5.2   A typical flow chart

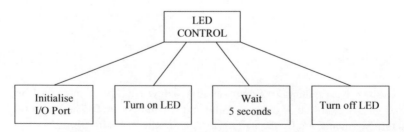

Figure 5.3   Example sequencing using structure charts

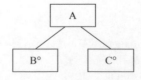

Figure 5.4   Example selection using structure charts

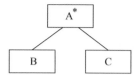

Figure 5.5   Example iteration using structure charts

5.2.2.3 Iteration

Iteration is shown by placing an asterisk sign at the top right-hand side of a rectangle. An example is given in Figure 5.5 where processes B and C are repeated.

5.2.2.4 Invoking Modules

In structure charts modules can be shown with double-sided rectangles. An example is shown in Figure 5.6 where module ADD is called.

Example 5.1

Draw the structure chart for an application where three numbers are read from the keyboard into a main program, their sum calculated using a module called SUM, and the result displayed by the main program.

Solution

The structure chart for this example is shown in Figure 5.7.

5.2.3 Pseudocode

One of the disadvantages of graphical design methods such as flow diagrams and structure charts is that it can take a long time to draw them and that it is not easy to modify them.

Pseudocode is a kind of structured English for describing the operation of algorithms. It allows the programmer to concentrate on the development of the algorithm independent

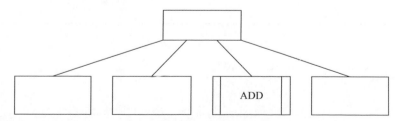

Figure 5.6   Invoking a module

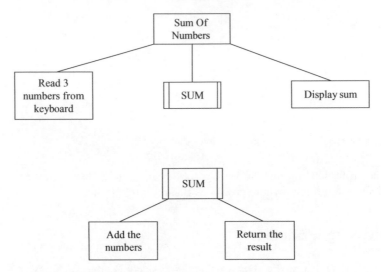

Figure 5.7   Structure chart for Example 5.1

of the details of the target language. There are no fixed rules or standards for developing pseudocode, and individual designers may have their own personal style of pseudocode. There are, however, guidelines to help the designer develop readable and powerful pseudocode. Pseudocode is based on the concept that any program consists of three major items: *sequencing*, *selection*, and *iteration*. Pseudocode is then developed using English sentences to describe algorithms, and this code cannot be compiled. If a program consists of a number of modules called by the main program then each module should be described using pseudocode. A brief description of the verbs and sentences that can be used in pseudocode is given in the rest of this section.

5.2.3.1   BEGIN–END

This construct is used to declare the beginning and end of a program or module. Keywords such as ':MAIN' can be used before BEGIN to declare the beginning of the main program:

:MAIN
BEGIN
 . . .
 . . .
END

Alternatively, the module name can be used:

:ADD
BEGIN
 . . .
 . . .
END

As shown in these examples, the lines should be indented to make the algorithm easier to read.

5.2.3.2 Sequencing

A sequence is a linear progression where the tasks are performed sequentially one after the other. Each action should be written on a new line and all the actions should be aligned with the same indent. The following keywords can be used for the description of the algorithm:

Input:	READ, GET, OBTAIN
Output:	SEND, PRINT, DISPLAY, SHOW
Initialize:	SET, CLEAR, INITIALIZE
Compute:	ADD, CALCULATE, DETERMINE
Actions:	TURN ON, TURN OFF

For example:

:MAIN
BEGIN
  Read three numbers
  Calculate their sum
  Display the result
END

5.2.3.3 IF–THEN–ELSE–ENDIF

The keywords IF, THEN, ELSE and ENDIF can be used to indicate that a decision is to be made. The general format of this construct is:

IF condition **THEN**
  statement
  statement
ELSE
  statement
  statement
ENDIF

The ELSE keyword and the statements following it are optional. In the following example they are omitted:

IF grade > 90 **THEN**
  Letter = 'A'
ENDIF

If the condition is true, the statements following the THEN are executed, otherwise the statements following the ELSE are executed. For example:

IF temperature > 100 **THEN**
  Turn off heater
  Start the engine

ELSE
  Turn on heater
ENDIF

5.2.3.4   REPEAT–UNTIL

This construct is used to specify a loop where the test is performed at the end of the loop, i.e. the loop is executed at least once, perhaps many times, depending upon the condition at the end of the loop. The loop continues forever if the condition is not satisfied. The general format is:

REPEAT
  Statement
  Statement
  Statement
UNTIL condition

In the following example the statements inside the loop are executed five times:

Set cnt = 0
REPEAT
  Turn on LED
  Wait 1 s
  Turn off LED
  Wait 1 s
  Increment cnt
UNTIL cnt = 5

5.2.3.5   DO–WHILE

This construct is similar to REPEAT–UNTIL, but here the loop is executed while the condition is true. The condition is tested at the end of the loop. The general format is:

DO
  statement
  statement
  statement
WHILE condition

In the following example the statements inside the loop are executed five times:

Set cnt = 0
DO
  Turn on LED
  Wait 1 s
  Turn off LED
  Wait 1 s
  Increment cnt
WHILE cnt < 5

5.2.3.6   WHILE–WEND

This construct is similar to REPEAT–UNTIL, but here the loop may never be executed, depending on the condition. The condition is tested at the beginning of the loop. The general format is:

WHILE condition
  statement
  statement
  statement
WEND

In the following example the loop is never executed:

I = 0
WHILE I > 0
  Turn on LED
  Wait 3 s
WEND

In this next example the loop is executed 10 times:

I = 0
WHILE I < 10
  Turn on motor
  Wait 2 seconds
  Turn off motor
  Increment I
WEND

5.2.3.7   CASE–CASE ELSE–ENDCASE

The CASE construct is used for multi-way branch operations. An expression is selected and, based on the value of this expression, a number of mutually exclusive tests can be done and statements can be executed for each case. The general format of this construct is:

CASE expression **OF**
  condition1:
    statement
    statement
  condition2:
    statement
    statement
  condition3:
    statement
    statement
  . . .
  . . .

CASE ELSE
  Statement
  Statement
END CASE

If the expression is equal to *condition1*, the statements following *condition1* are executed, if the expression is equal to *condition2*, the statements following *condition2* are executed and so on. If the expression is not equal to any of the specified conditions then the statements following the *CASE ELSE* are executed.

In the following example the points obtained by a student are calculated based on the grade:

CASE grade **OF**
  A: points $= 10$
  B: points $= 8$
  C: points $= 6$
  D: points $= 4$
  CASE ELSE   points $= 0$
END CASE

Notice that the above CASE construct can be implemented using the IF–THEN–ELSE construct as follows:

IF grade $=$ A **THEN**
  points $= 10$
ELSE IF grade $=$ B **THEN**
  points $= 8$
ELSE IF grade $=$ C **THEN**
  points $= 6$
ELSE IF grade $=$ D **THEN**
  points $= 4$
ELSE
  points $= 0$
END IF

5.2.3.8 Invoking Modules

Modules can be called using the CALL keyword and then specifying the name of the module. It is useful if the input parameters to be passed to the module are specified when a module is called. Similarly, at the header of the module description the input and the output parameters of a module should be specified. An example is given below.

Example 5.2

Write the pseudocode for an application where three numbers are read from the keyboard into a main program, their sum calculated using a module called SUM, and the result displayed by the main program.

Solution

The pseudocode for the main program and the module are given in Figure 5.8.

```
:MAIN
BEGIN
        Read 3 numbers a, b, c from the kayboard
        Call SUM (a, b, c)
        Display result
END


:SUM (I: a, b, c O: sum of numbers)
BEGIN
        Calculate the sum of a, b, c
        Return sum of numbers
END
```

Figure 5.8   Pseudocode for Example 5.2

5.3   EXERCISE

1. What are the three major components of a flow chart? Explain the function of each component with an example.

2. Draw a flow chart for a simple sort algorithm.

3. Draw a flow chart for a binary search algorithm.

4. What are the differences between a flow chart and a structure chart?

5. What are the three major components of a structure chart? Explain the function of each component with an example.

6. Draw a flow chart to show how a quadratic equation can be solved.

7. What are the advantages of pseudocode?

8. What are the basic components of pseudocode?

9. Write pseudocode to read the base and the height of a triangle from the keyboard, call a module to calculate the area of the triangle and display the area in the main program.

10. Explain how iteration can be done in pseudocode. Give an example.

11. Give an example of pseudocode to show how multi-way selection can be done using the CASE construct. Write the equivalent IF–ELSE–ENDIF construct.

FURTHER READING

[Alford, 1977]   Alford, M.W. A requirements engineering methodology for realtime processing requirements *IEEE Trans. Software Eng.*, SE-3, 1977, pp. 60–69.

[Baker, and Scallon, 1986]   Baker, T.P. and Scallon, G.M. An architecture for real-time software systems *IEEE Trans. Software Mag.*, **3**, 3, 1986, pp. 50–58.

[Bell et al., 1992]   Bell, G., Morrey, I., and Pugh, J. Software Engineering. Prentice Hall, Englewood Cliffs, NJ, 1992.

[Bennett, 1994]   Bennett, S. Real-Time Computer Control: An Introduction. Prentice Hall, Englewood Cliffs, NJ, 1994.

[Bibbero, 1977]   Bibbero, R.J. Microprocessors in Instruments and Control. John Wiley & Sons, Inc., New York, 1977.

6

Sampled Data Systems and the *z*-Transform

A sampled data system operates on discrete-time rather than continuous-time signals. A digital computer is used as the controller in such a system. A D/A converter is usually connected to the output of the computer to drive the plant. We will assume that all the signals enter and leave the computer at the same fixed times, known as the sampling times.

A typical sampled data control system is shown in Figure 6.1. The digital computer performs the controller or the compensation function within the system. The A/D converter converts the error signal, which is a continuous signal, into digital form so that it can be processed by the computer. At the computer output the D/A converter converts the digital output of the computer into a form which can be used to drive the plant.

6.1 THE SAMPLING PROCESS

A sampler is basically a switch that closes every T seconds, as shown in Figure 6.2. When a continuous signal $r(t)$ is sampled at regular intervals T, the resulting discrete-time signal is shown in Figure 6.3, where q represents the amount of time the switch is closed.

In practice the closure time q is much smaller than the sampling time T, and the pulses can be approximated by flat-topped rectangles as shown in Figure 6.4.

In control applications the switch closure time q is much smaller than the sampling time T and can be neglected. This leads to the ideal sampler with output as shown in Figure 6.5.

The ideal sampling process can be considered as the multiplication of a pulse train with a continuous signal, i.e.

$$r^*(t) = P(t)r(t), \tag{6.1}$$

where $P(t)$ is the delta pulse train as shown in Figure 6.6, expressed as

$$P(t) = \sum_{n=-\infty}^{\infty} \delta(t - nT); \tag{6.2}$$

thus,

$$r^*(t) = r(t) \sum_{n=-\infty}^{\infty} \delta(t - nT) \tag{6.3}$$

Microcontroller Based Applied Digital Control   D. Ibrahim
© 2006 John Wiley & Sons, Ltd

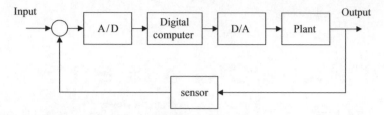

Figure 6.1   Sampled data control system

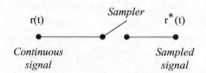

Figure 6.2   A sampler

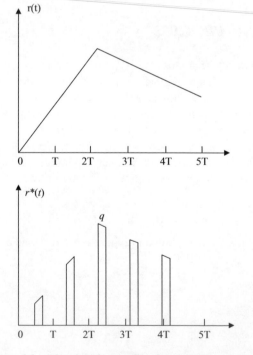

Figure 6.3   The signal $r(t)$ after the sampling operation

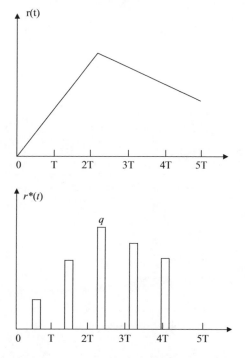

Figure 6.4   Sampled signal with flat-topped pulses

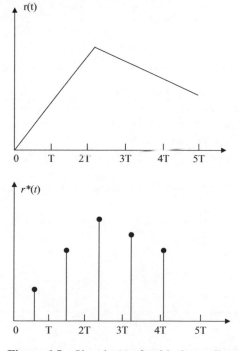

Figure 6.5   Signal $r(t)$ after ideal sampling

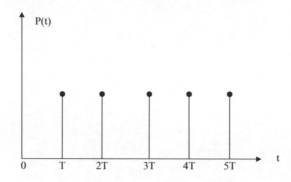

Figure 6.6   Delta pulse train

or

$$r^*(t) = \sum_{n=-\infty}^{\infty} r(nT)\delta(t - nT).$$ (6.4)

Now

$$r(t) = 0, \quad \text{for } t < 0,$$ (6.5)

and

$$r^*(t) = \sum_{n=0}^{\infty} r(nT)\delta(t - nT).$$ (6.6)

Taking the Laplace transform of (6.6) gives

$$R^*(s) = \sum_{n=0}^{\infty} r(nT)e^{-snT}.$$ (6.7)

Equation (6.7) represents the Laplace transform of a sampled continuous signal $r(t)$.

A D/A converter converts the sampled signal $r^*(t)$ into a continuous signal $y(t)$. The D/A can be approximated by a zero-order hold (ZOH) circuit as shown in Figure 6.7. This circuit remembers the last information until a new sample is obtained, i.e. the zero-order hold takes the value $r(nT)$ and holds it constant for $nT \leq t < (n + 1)T$, and the value $r(nT)$ is used during the sampling period.

The impulse response of a zero-order hold is shown in Figure 6.8. The transfer function of a zero-order hold is given by

$$G(t) = H(t) - H(t - T),$$ (6.8)

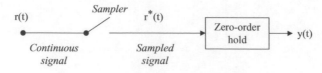

Figure 6.7   A sampler and zero-order hold

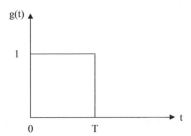

Figure 6.8   Impulse response of a zero-order hold

where $H(t)$ is the step function, and taking the Laplace transform yields

$$G(s) = \frac{1}{s} - \frac{e^{-Ts}}{s} = \frac{1 - e^{-Ts}}{s}. \tag{6.9}$$

A sampler and zero-order hold can accurately follow the input signal if the sampling time T is small compared to the transient changes in the signal. The response of a sampler and a zero-order hold to a ramp input is shown in Figure 6.9 for two different values of sampling period.

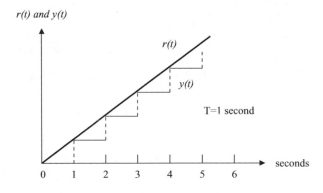

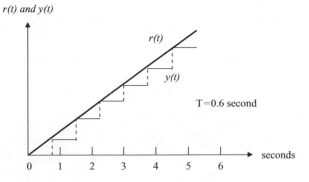

Figure 6.9   Response of a sampler and a zero-order hold for a ramp input

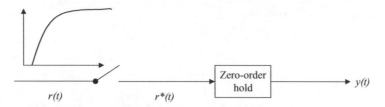

Figure 6.10   Ideal sampler and zero-order hold for Example 6.1

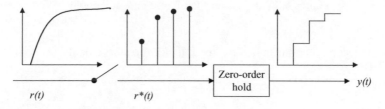

Figure 6.11   Solution for Example 6.1

Example 6.1

Figure 6.10 shows an ideal sampler followed by a zero-order hold. Assuming the input signal $r(t)$ is as shown in the figure, show the waveforms after the sampler and also after the zero-order hold.

Solution

The signals after the ideal sampler and the zero-order hold are shown in Figure 6.11.

6.2 THE z-TRANSFORM

Equation (6.7) defines an infinite series with powers of e^{-snT}. The z-transform is defined so that

$$Z = e^{sT};\tag{6.10}$$

the z-transform of the function $r(t)$ is $Z[r(t)] = R(z)$ which, from (6.7), is given by

$$R(z) = \sum_{n=0}^{\infty} r(nT)z^{-n}.\tag{6.11}$$

Notice that the z-transform consists of an infinite series in the complex variable z, and

$$R(z) = r(0) + r(T)z^{-1} + r(2T)z^{-2} + r(3T)z^{-3} + \ldots,$$

i.e. the $r(nT)$ are the coefficients of this power series at different sampling instants.

The z-transformation is used in sampled data systems just as the Laplace transformation is used in continuous-time systems. The response of a sampled data system can be determined easily by finding the z-transform of the output and then calculating the inverse z-transform,

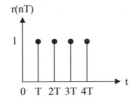

Figure 6.12   Unit step function

just like the Laplace transform techniques used in continuous-time systems. We will now look at how we can find the z-transforms of some commonly used functions.

6.2.1   Unit Step Function

Consider a unit step function as shown in Figure 6.12, defined as

$$r(nT) = \begin{cases} 0, & n < 0, \\ 1, & n \geq 0. \end{cases}$$

From (6.11),

$$R(z) = \sum_{n=0}^{\infty} r(nT)z^{-n} = \sum_{n=0}^{\infty} z^{-n} = 1 + z^{-1} + z^{-2} + z^{-3} + z^{-4} + \dots$$

or

$$R(z) = \frac{z}{z-1}, \quad \text{for } |z| > 1.$$

6.2.2   Unit Ramp Function

Consider a unit ramp function as shown in Figure 6.13, defined by

$$r(nT) = \begin{cases} 0, & n < 0, \\ nT, & n \geq 0. \end{cases}$$

From (6.11),

$$R(z) = \sum_{n=0}^{\infty} r(nT)z^{-n} = \sum_{n=0}^{\infty} nTz^{-n} = Tz^{-1} + 2Tz^{-2} + 3Tz^{-3} + 4Tz^{-4} + \dots$$

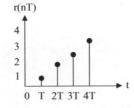

Figure 6.13   Unit ramp function

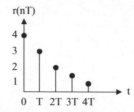

Figure 6.14   Exponential function

or

$$R(z) = \frac{Tz}{(z-1)^2}, \qquad \text{for } |z| > 1.$$

6.2.3   Exponential Function

Consider the exponential function shown in Figure 6.14, defined as

$$r(nT) = \begin{cases} 0, & n < 0, \\ e^{-anT}, & n \geq 0. \end{cases}$$

From (6.11),

$$R(z) = \sum_{n=0}^{\infty} r(nT)z^{-n} = \sum_{n=0}^{\infty} e^{-anT}z^{-n} = 1 + e^{-aT}z^{-1} + e^{-2aT}z^{-2} + e^{-3aT}z^{-3} + \dots$$

or

$$R(z) = \frac{1}{1 - e^{-aT}z^{-1}} = \frac{z}{z - e^{-aT}}, \qquad \text{for } |z| < e^{-aT}. \qquad (6.12)$$

6.2.4   General Exponential Function

Consider the general exponential function

$$r(n) = \begin{cases} 0, & n < 0, \\ p^n, & n \geq 0. \end{cases}$$

From (6.11),

$$R(z) = \sum_{n=0}^{\infty} r(nT)z^{-n} = \sum_{n=0}^{\infty} p^n z^{-n} = 1 + pz^{-1} + p^2z^{-2} + p^3z^{-3} + \dots$$

or

$$R(z) = \frac{z}{z - p}, \qquad \text{for } |z| < |p|.$$

Similarly, we can show that

$$R(p^{-k}) = \frac{z}{z - p^{-1}}.$$

6.2.5 Sine Function

Consider the sine function, defined as

$$r(nT) = \begin{cases} 0, & n < 0, \\ \sin n\omega T. & n \geq 0. \end{cases}$$

Recall that

$$\sin x = \frac{e^{jx} - e^{-jx}}{2j},$$

so that

$$r(nT) = \frac{e^{jn\omega T} - e^{-jn\omega T}}{2j} = \frac{e^{jn\omega T}}{2j} - \frac{e^{-jn\omega T}}{2j}. \tag{6.13}$$

But we already know from (6.12) that the z-transform of an exponential function is

$$R(e^{-anT}) = R(z) = \frac{z}{z - e^{-aT}}.$$

Therefore, substituting in (6.13) gives

$$R(z) = \frac{1}{2j}\left(\frac{z}{z - e^{j\omega T}} - \frac{z}{z - e^{-j\omega T}}\right) = \frac{1}{2j}\left(\frac{z(e^{j\omega T} - e^{-j\omega T})}{z^2 - z(e^{j\omega T} + e^{-j\omega T}) + 1}\right)$$

or

$$R(z) = \frac{z \sin \omega T}{z^2 - 2z \cos \omega T + 1}.$$

6.2.6 Cosine Function

Consider the cosine function, defined as

$$r(nT) = \begin{cases} 0, & n < 0, \\ \cos n\omega T, & n \geq 0. \end{cases}$$

Recall that

$$\cos x = \frac{e^{jx} + e^{-jx}}{2},$$

so that

$$r(nT) = \frac{e^{jn\omega T} + e^{-jn\omega T}}{2} = \frac{e^{jn\omega T}}{2} + \frac{e^{-jn\omega T}}{2}. \tag{6.14}$$

But we already know from (6.12) that the z-transform of an exponential function is

$$R(e^{-anT}) = R(z) = \frac{z}{z - e^{-aT}}.$$

Therefore, substituting in (6.14) gives

$$R(z) = \frac{1}{2}\left(\frac{z}{z - e^{j\omega T}} + \frac{z}{z - e^{-j\omega T}}\right)$$

or

$$R(z) = \frac{z(z - \cos \omega T)}{z^2 - 2z \cos \omega T + 1}.$$

6.2.7 Discrete Impulse Function

Consider the discrete impulse function defined as

$$\delta(n) = \begin{cases} 1, & n = 0, \\ 0, & n \neq 0. \end{cases}$$

From (6.11),

$$R(z) = \sum_{n=0}^{\infty} r(nT)z^{-n} = \sum_{n=0}^{\infty} z^{-n} = 1.$$

6.2.8 Delayed Discrete Impulse Function

The delayed discrete impulse function is defined as

$$\delta(n-k) = \begin{cases} 1, & n = k > 0, \\ 0, & n \neq k. \end{cases}$$

From (6.11),

$$R(z) = \sum_{n=0}^{\infty} r(nT)z^{-n} = \sum_{n=0}^{\infty} z^{-n} = z^{-n}.$$

6.2.9 Tables of z-Transforms

A table of z-transforms for the commonly used functions is given in Table 6.1 (a bigger table is given in Appendix A). As with the Laplace transforms, we are interested in the output response $y(t)$ of a system and we must find the inverse z-transform to obtain $y(t)$ from $Y(z)$.

6.2.10 The z-Transform of a Function Expressed as a Laplace Transform

It is important to realize that although we denote the z-transform equivalent of $G(s)$ by $G(z)$, $G(z)$ is *not* obtained by simply substituting z for s in $G(s)$. We can use one of the following methods to find the z-transform of a function expressed in Laplace transform format:

- Given $G(s)$, calculate the time response $g(t)$ by finding the inverse Laplace transform of $G(s)$. Then find the z-transform either from the first principles, or by looking at the z-transform tables.

- Given $G(s)$, find the z-tranform $G(z)$ by looking at the tables which give the Laplace transforms and their equivalent z-transforms (e.g. Table 6.1).

- Given the Laplace transform $G(s)$, express it in the form $G(s) = N(s)/D(s)$ and then use the following formula to find the z-transform $G(z)$:

$$G(z) = \sum_{n=1}^{p} \frac{N(x_n)}{D'(x_n)} \frac{1}{1 - e^{x_n T}z^{-1}}, \tag{6.15}$$

Table 6.1   Some commonly used z-transforms

$f(kT)$	$F(z)$
$\delta(t)$	1
1	$\dfrac{z}{z-1}$
kT	$\dfrac{Tz}{(z-1)^2}$
e^{-akT}	$\dfrac{z}{z-e^{-aT}}$
kTe^{-akT}	$\dfrac{Tze^{-aT}}{(z-e^{-aT})^2}$
a^k	$\dfrac{z}{z-a}$
$1-e^{-akT}$	$\dfrac{z(1-e^{-aT})}{(z-1)(z-e^{-aT})}$
$\sin akT$	$\dfrac{z\sin aT}{z^2-2z\cos aT+1}$
$\cos akT$	$\dfrac{z(z-\cos aT)}{z^2-2z\cos aT+1}$

where $D' = \partial D/\partial s$ and the $x_n, n = 1, 2, \ldots, p$, are the roots of the equation $D(s) = 0$. Some examples are given below.

Example 6.2

Let

$$G(s) = \frac{1}{s^2 + 5s + 6}.$$

Determine $G(z)$ by the methods described above.

Solution

Method 1: By finding the inverse Laplace transform. We can express $G(s)$ as a sum of its partial fractions:

$$G(s) = \frac{1}{(s+3)(s+2)} = \frac{1}{s+2} - \frac{1}{s+3}. \tag{6.16}$$

The inverse Laplace transform of (6.16) is

$$g(t) = L^{-1}[G(s)] = e^{-2t} - e^{-3t}. \tag{6.17}$$

From the definition of the z-transforms we can write (6.17) as

$$G(z) = \sum_{n=0}^{\infty}(e^{-2nT} - e^{-3nT})z^{-n}$$

$$= (1 + e^{-2T}z^{-1} + e^{-4T}z^{-2} + \ldots) - (1 + e^{-3T}z^{-1} + e^{-6T}z^{-2} + \ldots)$$

$$= \frac{z}{z - e^{-2T}} - \frac{z}{z - e^{-3T}}$$

or

$$G(z) = \frac{z(e^{-2T} - e^{-3T})}{(z - e^{-2T})(z - e^{-3T})}.$$

Method 2: By using the z-transform transform tables for the partial product. From Table 6.1, the z-transform of $1/(s+a)$ is $z/(z - e^{-aT})$. Therefore the z-transform of (6.16) is

$$G(z) = \frac{z}{z - e^{-2T}} - \frac{z}{z - e^{-3T}}$$

or

$$G(z) = \frac{z(e^{-2T} - e^{-3T})}{(z - e^{-2T})(z - e^{-3T})}.$$

Method 3: By using the z-transform tables for G(s). From Table 6.1, the z-transform of

$$G(s) = \frac{b - a}{(s + a)(s + b)} \tag{6.18}$$

is

$$G(z) = \frac{z(e^{-aT} - e^{-bT})}{(z - e^{-aT})(z - e^{-bT})}. \tag{6.19}$$

Comparing (6.18) with (6.16) we have, $a = 2$, $b = 3$. Thus, in (6.19) we get

$$G(z) = \frac{z(e^{-2T} - e^{-3T})}{(z - e^{-2T})(z - e^{-3T})}.$$

Method 4: By using equation (6.15). Comparing our expression

$$G(s) = \frac{1}{s^2 + 5s + 6}$$

with (6.15), we have $N(s) = 1$, $D(s) = s^2 + 5s + 6$ and $D'(s) = 2s + 5$, and the roots of $D(s) = 0$ are $x_1 = -2$ and $x_2 = -3$. Using (6.15),

$$G(z) = \sum_{n=1}^{2} \frac{N(x_n)}{D'(x_n)} \frac{1}{1 - e^{x_n T} z^{-1}}$$

or, when $x_1 = -2$,

$$G_1(z) = \frac{1}{1} \frac{1}{1 - e^{-2T} z^{-1}}$$

and when $x_1 = -3$,

$$G_2(z) = \frac{1}{-1} \frac{1}{1 - e^{-3T} z^{-1}}.$$

Thus,

$$G(z) = \frac{1}{1 - e^{-2T} z^{-1}} - \frac{1}{1 - e^{-3T} z^{-1}} = \frac{z}{z - e^{-2T}} - \frac{z}{z - e^{-3T}}$$

or

$$G(z) = \frac{z(e^{-2T} - e^{-3T})}{(z - e^{-2T})(z - e^{-3T})}.$$

6.2.11 *Properties of z-Transforms*

Most of the properties of the z-transform are analogs of those of the Laplace transforms. Important z-transform properties are discussed in this section.

1. Linearity property
   Suppose that the z-transform of $f(nT)$ is $F(z)$ and the z-transform of $g(nT)$ is $G(z)$. Then

   $$Z[f(nT) \pm g(nT)] = Z[f(nT)] \pm Z[g(nT)] = F(z) \pm G(z) \qquad (6.20)$$

   and for any scalar a

   $$Z[af(nT)] = aZ[f(nT)] = aF(z) \qquad (6.21)$$

2. Left-shifting property
   Suppose that the z-transform of $f(nT)$ is $F(z)$ and let $y(nT) = f(nT + mT)$. Then

   $$Y(z) = z^m F(z) - \sum_{i=0}^{m-1} f(iT)z^{m-i}. \qquad (6.22)$$

   If the initial conditions are all zero, i.e. $f(iT) = 0$, $i = 0, 1, 2, \ldots, m - 1$, then,

   $$Z[f(nT + mT)] = z^m F(z). \qquad (6.23)$$

3. Right-shifting property
   Suppose that the z-transform of $f(nT)$ is $F(z)$ and let $y(nT) = f(nT - mT)$. Then

   $$Y(z) = z^{-m} F(z) + \sum_{i=0}^{m-1} f(iT - mT)z^{-i}. \qquad (6.24)$$

   If $f(nT) = 0$ for $k < 0$, then the theorem simplifies to

   $$Z[f(nT - mT)] = z^{-m} F(z). \qquad (6.25)$$

4. Attenuation property
   Suppose that the z-transform of $f(nT)$ is $F(z)$. Then,

   $$Z[e^{-anT} f(nT)] = F[ze^{aT}]. \qquad (6.26)$$

   This result states that if a function is multiplied by the exponential e^{-anT} then in the z-transform of this function z is replaced by ze^{aT}.

5. Initial value theorem
   Suppose that the z-transform of $f(nT)$ is $F(z)$. Then the initial value of the time response is given by

   $$\lim_{n \to 0} f(nT) = \lim_{z \to \infty} F(z). \qquad (6.27)$$

6. Final value theorem

Suppose that the z-transform of $f(nT)$ is $F(z)$. Then the final value of the time response is given by

$$\lim_{n \to \infty} f(nT) = \lim_{z \to 1}(1 - z^{-1})F(z).\qquad(6.28)$$

Note that this theorem is valid if the poles of $(1 - z^{-1})F(z)$ are inside the unit circle or at $z = 1$.

Example 6.3

The z-transform of a unit ramp function $r(nT)$ is

$$R(z) = \frac{Tz}{(z - 1)^2}.$$

Find the z-transform of the function $5r(nT)$.

Solution

Using the linearity property of z-transforms,

$$Z[5r(nT)] = 5R(z) = \frac{5Tz}{(z - 1)^2}.$$

Example 6.4

The z-transform of trigonometric function $r(nT) = \sin nwT$ is

$$R(z) = \frac{z \sin wT}{z^2 - 2z \cos wT + 1}.$$

find the z-transform of the function $y(nT) = e^{-2T} \sin nWT$.

Solution

Using property 4 of the z-transforms,

$$Z[y(nT)] = Z[e^{-2T}r(nT)] = R[ze^{2T}].$$

Thus,

$$Z[y(nT)] = \frac{ze^{2T} \sin wT}{(ze^{2T})^2 - 2ze^{2T} \cos wT + 1} = \frac{ze^{2T} \sin wT}{z^2 e^{4T} - 2ze^{2T} \cos wT + 1}$$

or, multiplying numerator and denominator by e^{-4T},

$$Z[y(nT)] = \frac{ze^{-2T} \sin wT}{z^2 - 2ze^{-2T} + e^{-4T}}.$$

Example 6.5

Given the function

$$G(z) = \frac{0.792z}{(z - 1)(z^2 - 0.416z + 0.208)},$$

find the final value of $g(nT)$.

Solution

Using the final value theorem,

$$\lim_{n \to \infty} g(nT) = \lim_{z \to 1} (1 - z^{-1}) \frac{0.792z}{(z - 1)(z^2 - 0.416z + 0.208)}$$

$$= \lim_{z \to 1} \frac{0.792}{z^2 - 0.416z + 0.208}$$

$$= \frac{0.792}{1 - 0.416 + 0.208} = 1.$$

6.2.12 Inverse z-Transforms

The inverse z-transform is obtained in a similar way to the inverse Laplace transforms. Generally, the z-transforms are the ratios of polynomials in the complex variable z, with the numerator polynomial being of order no higher than the denominator. By finding the inverse z-transform we find the sequence associated with the given z-transform polynomial. As in the case of inverse Laplace transforms, we are interested in the output time response of a system. Therefore, we use an inverse transform to obtain $y(t)$ from $Y(z)$. There are several methods to find the inverse z-transform of a given function. The following methods will be described here:

- power series (long division);

- expanding $Y(z)$ into partial fractions and using z-transform tables to find the inverse transforms;

- obtaining the inverse z-transform using an inversion integral.

Given a z-transform function $Y(z)$, we can find the coefficients of the associated sequence $y(nT)$ at the sampling instants by using the inverse z-transform. The time function $y(t)$ is then determined as

$$y(t) = \sum_{n=0}^{\infty} y(nT)\delta(t - nT).$$

Method 1: Power series. This method involves dividing the denominator of $Y(z)$ into the numerator such that a power series of the form

$$Y(z) = y_0 + y_1 z^{-1} + y_2 z^{-2} + y_3 z^{-3} + \ldots$$

is obtained. Notice that the values of $y(n)$ are the coefficients in the power series.

Example 6.6

Find the inverse z-transform for the polynomial

$$Y(z) = \frac{z^2 + z}{z^2 - 3z + 4}.$$

Solution

Dividing the denominator into the numerator gives

$$z^2 - 3z + 4 \overline{\smash{\big)}\ \begin{array}{l} 1 + 4z^{-1} + 8z^{-2} + 8z^{-3} \\ \hline z^2 + z \\ \underline{z^2 - 3z + 4} \\ 4z - 4 \\ \underline{4z - 12 + 16z^{-1}} \\ 8 - 16z^{-1} \\ \underline{8 - 24z^{-1} + 32z^{-2}} \\ 8z^{-1} - 32z^{-2} \\ \underline{8z^{-1} - 24z^{-2} + 32z^{-3}} \\ \cdots \end{array}}$$

and the coefficients of the power series are

$$\begin{aligned} y(0) &= 1, \\ y(T) &= 4, \\ y(2T) &= 8, \\ y(3T) &= 8, \\ &\cdots \end{aligned}$$

The required sequence is

$$y(t) = \delta(t) + 4\delta(t - T) + 8\delta(t - 2T) + 8\delta(t - 3T) + \cdots$$

Figure 6.15 shows the first few samples of the time sequence $y(nT)$.

Example 6.7

Find the inverse z-transform for $Y(z)$ given by the polynomial

$$Y(z) = \frac{z}{z^2 - 3z + 2}.$$

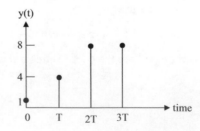

Figure 6.15   First few samples of $y(t)$

Solution

Dividing the denominator into the numerator gives

$$
z^2 - 3z + 2 \overline{\smash{\big)}\,
\begin{array}{l}
z^{-1} + 3z^{-2} + 7z^{-3} + 15z^{-4} \\
z \\
\underline{z - 3 + 2z^{-1}} \\
\quad 3 - 2z^{-1} \\
\quad \underline{3 - 9z^{-1} + 6z^{-2}} \\
\qquad 7z^{-1} - 6z^{-2} \\
\qquad \underline{7z^{-1} - 21z^{-2} + 14z^{-3}} \\
\qquad\quad 15z^{-2} - 14z^{-3} \\
\qquad\quad \underline{15z^{-2} - 45z^{-3} + 30z^{-4}} \\
\qquad\qquad \cdots
\end{array}}
$$

and the coefficients of the power series are

$$
\begin{aligned}
y(0) &= 0 \\
y(T) &= 1 \\
y(2T) &= 3 \\
y(3T) &= 7 \\
y(4T) &= 15 \\
&\cdots
\end{aligned}
$$

The required sequence is thus

$$
y(t) = \delta(t - T) + 3\delta(t - 2T) + 7\delta(t - 3T) + 15\delta(t - 4T) + \dots.
$$

Figure 6.16 shows the first few samples of the time sequence $y(nT)$.

The disadvantage of the power series method is that it does not give a closed form of the resulting sequence. We often need a closed-form result, and other methods should be used when this is the case.

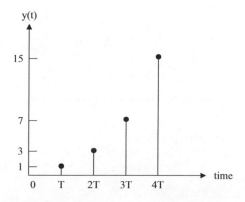

Figure 6.16 First few samples of $y(t)$

Method 2: Partial fractions. Similar to the inverse Laplace transform techniques, a partial fraction expansion of the function $Y(z)$ can be found, and then tables of known z-transforms can be used to determine the inverse z-transform. Looking at the z-transform tables, we see that there is usually a z term in the numerator. It is therefore more convenient to find the partial fractions of the function $Y(z)/z$ and then multiply the partial fractions by z to obtain a z term in the numerator.

Example 6.8

Find the inverse z-transform of the function

$$Y(z) = \frac{z}{(z-1)(z-2)}$$

Solution

The above expression can be written as

$$\frac{Y(z)}{z} = \frac{1}{(z-1)(z-2)} = \frac{A}{z-1} + \frac{B}{z-2}.$$

The values of A and B can be found by equating like powers in the numerator, i.e.

$$A(z-2) + B(z-1) \equiv 1.$$

We find $A = -1$, $B = 1$, giving

$$\frac{Y(z)}{z} = \frac{-1}{z-1} + \frac{1}{z-2}$$

or

$$Y(z) = \frac{-z}{z-1} + \frac{z}{z-2}$$

From the z-transform tables we find that

$$y(nT) = -1 + 2^n$$

and the coefficients of the power series are

$$
\begin{aligned}
y(0) &= 0, \\
y(T) &= 1, \\
y(2T) &= 3, \\
y(3T) &= 7, \\
y(4T) &= 15, \\
&\cdots
\end{aligned}
$$

so that the required sequence is

$$y(t) = \delta(t-T) + 3\delta(t-2T) + 7\delta(t-3T) + 15\delta(t-4T) + \ldots.$$

Example 6.9

Find the inverse z-transform of the function

$$Y(z) = \frac{1}{(z-1)(z-2)}.$$

Solution

The above expression can be written as

$$\frac{Y(z)}{z} = \frac{1}{z(z-1)(z-2)} = \frac{A}{z} + \frac{B}{z-1} + \frac{C}{z-2}.$$

The values of A, B and C can be found by equating like powers in the numerator, i.e.

$$A(z-1)(z-2) + Bz(z-2) + Cz(z-1) \equiv 1$$

or

$$A(z^2 - 3z + 2) + Bz^2 - 2Bz + Cz^2 - Cz \equiv 1,$$

giving

$$
\begin{aligned}
A + B + C &= 0, \\
-3A - 2B - C &= 0, \\
2A &= 1.
\end{aligned}
$$

The values of the coefficients are found to be $A = 0.5$, $B = -1$ and $C = 0.5$. Thus,

$$\frac{Y(z)}{z} = \frac{1}{2z} - \frac{1}{z-1} + \frac{1}{2(z-2)}$$

or

$$Y(z) = \frac{1}{2} - \frac{z}{z-1} + \frac{z}{2(z-2)}.$$

Using the inverse z-transform tables, we find

$$y(nT) = a - 1 + \frac{2^n}{2} = a - 1 + 2^{n-1}$$

where

$$a = \begin{cases} 1/2, & n = 0, \\ 0, & n \neq 0, \end{cases}$$

the coefficients of the power series are

$$
\begin{aligned}
y(0) &= 0 \\
y(T) &= 0 \\
y(2T) &= 1 \\
y(3T) &= 3 \\
y(4T) &= 7 \\
y(5T) &= 15 \\
&\cdots,
\end{aligned}
$$

and the required sequence is

$$y(t) = \delta(t - 2T) + 3\delta(t - 3T) + 7\delta(t - 4T) + 15\delta(t - 5T) + \cdots.$$

The process of finding inverse z-transforms is aided by considering what form is taken by the roots of $Y(z)$. It is useful to distinguish the case of distinct real roots and that of multiple order roots.

Case I: Distinct real roots. When $Y(z)$ has distinct real roots in the form

$$Y(z) = \frac{N(z)}{(z - p_1)(z - p_2)(z - p_3)\ldots(z - p_n)},$$

then the partial fraction expansion can be written as

$$Y(z) = \frac{A_1}{z - p_1} + \frac{A_2}{z - p_2} + \frac{A_3}{z - p_3} + \ldots + \frac{A_n}{z - p_n}$$

and the coefficients A_i can easily be found as

$$A_i = (z - p_i)\, Y(z)|_{z=p_i} \quad \text{for } i = 1,2,3,\ldots,n.$$

Example 6.10

Using the partial expansion method described above, find the inverse z-transform of

$$Y(z) = \frac{z}{(z - 1)(z - 2)}.$$

Solution

Rewriting the function as

$$\frac{Y(z)}{z} = \frac{A}{z - 1} + \frac{B}{z - 2},$$

we find that

$$A = (z - 1)\, \frac{1}{(z - 1)(z - 2)}\bigg|_{z=1} = -1,$$

$$B = (z - 2)\, \frac{1}{(z - 1)(z - 2)}\bigg|_{z=2} = 1.$$

Thus,

$$Y(z) = \frac{z}{z - 1} + \frac{z}{z - 2}$$

and the inverse z-transform is obtained from the tables as

$$y(nT) = -1 + 2^n,$$

which is the same answer as in Example 6.7.

Example 6.11

Using the partial expansion method described above, find the inverse z-transform of

$$Y(z) = \frac{z^2 + z}{(z - 0.5)(z - 0.8)(z - 1)}.$$

Solution

Rewriting the function as

$$\frac{Y(z)}{z} = \frac{A}{z - 0.5} + \frac{B}{z - 0.8} + \frac{C}{z - 1}$$

we find that

$$A = (z - 0.5) \left. \frac{z + 1}{(z - 0.5)(z - 0.8)(z - 1)} \right|_{z=0.5} = 10,$$

$$B = (z - 0.8) \left. \frac{z + 1}{(z - 0.5)(z - 0.8)(z - 1)} \right|_{z=0.8} = -30,$$

$$C = (z - 1) \left. \frac{z + 1}{(z - 0.5)(z - 0.8)(z - 1)} \right|_{z=1} = 20.$$

Thus,

$$Y(z) = \frac{10z}{z - 0.5} - \frac{30z}{z - 0.8} + \frac{20z}{z - 1}$$

The inverse transform is found from the tables as

$$y(nT) = 10(0.5)^n - 30(0.8)^n + 20$$

The coefficients of the power series are

$$\begin{aligned} y(0) &= 0 \\ y(T) &= 1 \\ y(2T) &= 3.3 \\ y(3T) &= 5.89 \end{aligned}$$

$$\ldots$$

and the required sequence is

$$y(t) = \delta(t - T) + 3.3\delta(t - 2T) + 5.89\delta(t - 3T) + \ldots.$$

Case II: Multiple order roots. When $Y(z)$ has multiple order roots of the form

$$Y(z) = \frac{N(z)}{(z - p_1)(z - p_1)^2(z - p_1)^3 \ldots (z - p_1)^r},$$

then the partial fraction expansion can be written as

$$Y(z) = \frac{\lambda_1}{z - p_1} + \frac{\lambda_2}{(z - p_2)^2} + \frac{\lambda_3}{(z - p_1)^3} + \ldots + \frac{\lambda_r}{(z - p_1)^r}$$

and the coefficients λ_i can easily be found as

$$\lambda_{r-k} = \frac{1}{k!} \left. \frac{d^k}{dz^k} [(z - p_i)^r (X(z)/z)] \right|_{z=p_i}. \tag{6.29}$$

Example 6.12

Using (6.29), find the inverse z-transform of

$$Y(z) = \frac{z^2 + 3z - 2}{(z + 5)(z - 0.8)(z - 2)^2}.$$

Solution

Rewriting the function as

$$\frac{Y(z)}{z} = \frac{z^2 + 3z - 2}{z(z + 5)(z - 0.8)(z - 2)^2} = \frac{A}{z} + \frac{B}{z + 5} + \frac{C}{z - 0.8} + \frac{D}{(z - 2)} + \frac{E}{(z - 2)^2}$$

we obtain

$$A = z \left. \frac{z^2 + 3z - 2}{z(z + 5)(z - 0.8)(z - 2)^2} \right|_{z=0} = \frac{-2}{5 \times (-0.8) \times 4} = 0.125,$$

$$B = (z + 5) \left. \frac{z^2 + 3z - 2}{z(z + 5)(z - 0.8)(z - 2)^2} \right|_{z=-5} = \frac{8}{-5 \times (-5.8) \times 49} = 0.0056,$$

$$C = (z - 0.8) \left. \frac{z^2 + 3z - 2}{z(z + 5)(z - 0.8)(z - 2)^2} \right|_{z=0.8} = \frac{1.04}{0.8 \times 5.8 \times 1.14} = 0.16,$$

$$E = (z - 2)^2 \left. \frac{z^2 + 3z - 2}{z(z + 5)(z - 0.8)(z - 2)^2} \right|_{z=2} = \frac{8}{2 \times 7 \times 1.2} = 0.48,$$

$$D = \frac{d}{dz} \left[\frac{z^2 + 3z - 2}{z(z + 5)(z - 0.8)} \right] \Bigg|_{z=2}$$

$$= \frac{\left[z(z + 5)(z - 0.8(2z + 3) - (z^2 + 3z - 2)(3z^2 + 8.4z - 4) \right]}{(z^3 + 4.2z^2 - 4z)^2} \Bigg|_{z=2} = -0.29.$$

We can now write $Y(z)$ as

$$Y(z) = 0.125 + \frac{0.0056z}{z + 5} + \frac{0.016z}{z - 0.8} - \frac{0.29z}{(z - 2)} + \frac{0.48z}{(z - 2)^2}$$

The inverse transform is found from the tables as

$$y(nT) = 0.125a + 0.0056(-5)^n + 0.016(0.8)^n - 0.29(2)^n + 0.24n(2)^n,$$

where

$$a = \begin{cases} 1, & n = 0, \\ 0, & n \neq 0. \end{cases}$$

Method 3: Inversion formula method. The inverse z-transform can be obtained using the inversion integral, defined by

$$y(nT) = \frac{1}{2\pi j} \oint_r Y(z) z^{n-1} dz. \qquad (6.30)$$

Using the theorem of residues, the above integral can be evaluated via the expression

$$y(nT) = \sum_{\substack{\text{at poles of} \\ [Y(z)z^{k-1}]}} [\text{residues of } Y(z)z^{n-1}]. \tag{6.31}$$

If the function has a simple pole at $z = a$, then the residue is evaluated as

$$[\text{residue}]|_{z=a} = [(z - a)Y(z)z^{n-1}]. \tag{6.32}$$

Example 6.13

Using the inversion formula method, find the inverse z-transform of

$$Y(z) = \frac{z}{(z - 1)(z - 2)}.$$

Solution

Using (6.31) and (6.32):

$$y(nT) = \frac{z^n}{z - 2}\bigg|_{z=1} + \frac{z^n}{z - 1}\bigg|_{z=2} = -1 + 2^n$$

which is the same answer as in Example 6.9.

Example 6.14

Using the inversion formula method, find the inverse z-transform of

$$Y(z) = \frac{z}{(z - 1)(z - 2)(z - 3)}.$$

Solution

Using (6.31) and (6.32),

$$y(nT) = \frac{z^n}{(z - 2)(z - 3)}\bigg|_{z=1} + \frac{z^n}{(z - 1)(z - 3)}\bigg|_{z=2} + \frac{z^n}{(z - 1)(z - 2)}\bigg|_{z=3} = \frac{1}{2} - 2^n + \frac{3^n}{2}.$$

6.3 PULSE TRANSFER FUNCTION AND MANIPULATION OF BLOCK DIAGRAMS

The pulse transfer function is the ratio of the z-transform of the sampled output and the input at the sampling instants.

Suppose we wish to sample a system with output response given by

$$y(s) = e^*(s)G(s). \tag{6.33}$$

Figure 6.17   Sampling a system

as illustrated in Figure 6.17. We sample the output signal to obtain

$$y^*(s) = [e^*(s)G(s)]^* = e^*(s)G^*(s) \tag{6.34}$$

and

$$y(z) = e(z)G(z). \tag{6.35}$$

Equations (6.34) and (6.35) tell us that if at least one of the continuous functions has been sampled, then the z-transform of the product is equal to the product of the z-transforms of each function (note that $[e^*(s)]^* = [e^*(s)]$, since sampling an already sampled signal has no further effect). $G(z)$ is the transfer function between the sampled input and the output at the sampling instants and is called the *pulse transfer function*. Notice from (6.35) that we have no information about the output $y(z)$ between the sampling instants.

6.3.1 Open-Loop Systems

Some examples of manipulating open-loop block diagrams are given in this section.

Example 6.15

Figure 6.18 shows an open-loop sampled data system. Derive an expression for the z-transform of the output of the system.

Solution

For this system we can write

$$y(s) = e^*(s)KG(s)$$

or

$$y^*(s) = [e^*(s)KG(s)]^* = e^*(s)KG^*(s)$$

and

$$y(z) = e(z)KG(z).$$

Example 6.16

Figure 6.19 shows an open-loop sampled data system. Derive an expression for the z-transform of the output of the system.

Figure 6.18   Open-loop system

Figure 6.19   Open-loop system

Solution

The following expressions can be written for the system:

$$y(s) = e^*(s)G_1(s)G_2(s)$$

or

$$y^*(s) = [e^*(s)G_1(s)G_2(s)]^* = e^*(s)[G_1G_2]^*(s)$$

and

$$y(z) = e(z)G_1G_2(z),$$

where

$$G_1G_2(z) = Z\{G_1(s)G_2(s)\} \neq G_1(z)G_2(z).$$

For example, if

$$G_1(s) = \frac{1}{s}$$

and

$$G_2(s) = \frac{a}{s+a},$$

then from the z-transform tables,

$$Z\{G_1(s)G_2(s)\} = Z\left\{\frac{a}{s(s+a)}\right\} = \frac{z(1 - e^{-aT})}{(z-1)(z - e^{-aT})}$$

and the output is given by

$$y(z) = e(z)\frac{z(1 - e^{-aT})}{(z-1)(z - e^{-aT})}.$$

Example 6.17

Figure 6.20 shows an open-loop sampled data system. Derive an expression for the z-transform of the output of the system.

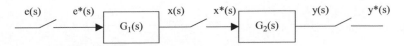

Figure 6.20   Open-loop system

Solution

The following expressions can be written for the system:

$$x(s) = e^*(s)G_1(s)$$

or

$$x^*(s) = e^*(s)G_1^*(s), \tag{6.36}$$

and

$$y(s) = x^*(s)G_2(s)$$

or

$$y^*(s) = x^*(s)G_2^*(s). \tag{6.37}$$

From (6.37) and (6.38),

$$y^*(s) = e^*(s)G_1^*(s)G_2^*(s),$$

which gives

$$y(z) = e(z)G_1(z)G_2(z).$$

For example, if

$$G_1(s) = \frac{1}{s} \quad \text{and} \quad G_2(s) = \frac{a}{s+a},$$

then

$$Z\{G_1(s)\} = \frac{z}{z-1} \quad \text{and} \quad Z\{G_2(s)\} = \frac{az}{z - ze^{-aT}},$$

and the output function is given by

$$y(z) = e(z)\frac{z}{z-1}\frac{az}{z - ze^{-aT}}$$

or

$$y(z) = e(z)\frac{az}{(z-1)(1 - e^{-aT})}.$$

6.3.2 Open-Loop Time Response

The open-loop time response of a sampled data system can be obtained by finding the inverse z-transform of the output function. Some examples are given below.

Example 6.18

A unit step signal is applied to the electrical RC system shown in Figure 6.21. Calculate and draw the output response of the system, assuming a sampling period of $T = 1$ s.

Solution

The transfer function of the RC system is

$$G(s) = \frac{1}{s+1}.$$

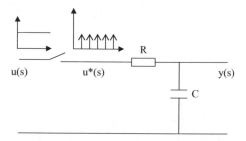

Figure 6.21   RC system with unit step input

For this system we can write

$$y(s) = u^*(s)G(s)$$

and

$$y^*(s) = u^*(s)G^*(s),$$

and taking z-transforms gives

$$y(z) = u(z)G(z).$$

The z-transform of a unit step function is

$$u(z) = \frac{z}{z-1}$$

and the z-transform of $G(s)$ is

$$G(z) = \frac{z}{z-e^{-T}}.$$

Thus, the output z-transform is given by

$$y(z) = u(z)G(z) = \frac{z}{z-1}\frac{z}{z-e^{-T}} = \frac{z^2}{(z-1)(z-e^{-T})};$$

since $T = 1$ s and $e^{-1} = 0.368$, we get

$$y(z) = \frac{z^2}{(z-1)(z-0.368)}.$$

The output response can be obtained by finding the inverse z-transform of $y(z)$. Using partial fractions,

$$\frac{y(z)}{z} = \frac{A}{z-1} + \frac{B}{z-0.368}.$$

Calculating A and B, we find that

$$\frac{y(z)}{z} = \frac{1.582}{z-1} - \frac{0.582}{z-0.368}$$

or

$$y(z) = \frac{1.582z}{z-1} - \frac{0.582z}{z-0.368}.$$

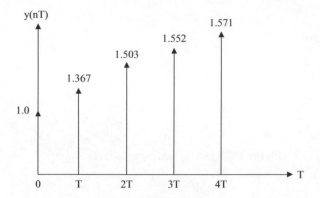

Figure 6.22   RC system output response

From the z-transform tables we find

$$y(nT) = 1.582 - 0.582(0.368)^n.$$

The first few output samples are

$$y(0) = 1,$$
$$y(1) = 1.367,$$
$$y(2) = 1.503,$$
$$y(3) = 1.552,$$
$$y(4) = 1.571,$$

and the output response (shown in Figure 6.22) is given by

$$y(nT) = \delta(T) + 1.367\delta(t - T) + 1.503\delta(t - 2T) + 1.552\delta(t - 3T) + 1.571\delta(t - 4T) + \ldots.$$

It is important to notice that the response is only known at the sampling instants. For example, in Figure 6.22 the capacitor discharges through the resistor between the sampling instants, and this causes an exponential decay in the response between the sampling intervals. But this behaviour between the sampling instants cannot be determined by the z-transform method of analysis.

Example 6.19

Assume that the system in Example 6.17 is used with a zero-order hold (see Figure 6.23). What will the system output response be if (i) a unit step input is applied, and (ii) if a unit ramp input is applied.

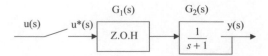

Figure 6.23   RC system with a zero-order hold

Solution

The transfer function of the zero-order hold is

$$G_1(s) = \frac{1 - e^{-Ts}}{s}$$

and that of the RC system is

$$G(s) = \frac{1}{s+1}.$$

For this system we can write

$$y(s) = u^*(s)G_1G_2(s)$$

and

$$y^*(s) = u^*(s)[G_1G_2]^*(s)$$

or, taking z-transforms,

$$y(z) = u(z)G_1G_2(z).$$

Now, $T = 1$ s and

$$G_1G_2(s) = \frac{1 - e^{-s}}{s} \frac{1}{s+1},$$

and by partial fraction expansion we can write

$$G_1G_2(s) = (1 - e^{-s})\left(\frac{1}{s} - \frac{1}{s+1}\right).$$

From the z-transform tables we then find that

$$G_1G_2(z) = (1 - z^{-1})\left(\frac{z}{z-1} - \frac{z}{z-e^{-1}}\right) = \frac{0.63}{z - 0.37}.$$

(i) For a unit step input,

$$u(z) = \frac{z}{z-1}$$

and the system output response is given by

$$y(z) = \frac{0.63z}{(z-1)(z-0.37)}.$$

Using the partial fractions method, we can write

$$\frac{y(z)}{z} = \frac{A}{z-1} + \frac{B}{z-0.37},$$

where $A = 1$ and $B = -1$; thus,

$$y(z) = \frac{z}{z-1} - \frac{z}{z-0.37}.$$

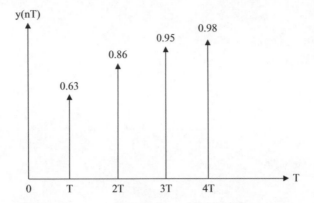

Figure 6.24   Step input time response of Example 6.19

From the inverse z-transform tables we find that the time response is given by

$$y(nT) = a - (0.37)^n,$$

where a is the unit step function; thus

$$y(nT) = 0.63\delta(t - 1) + 0.86\delta(t - 2) + 0.95\delta(t - 3) + 0.98\delta(t - 4) + \ldots.$$

The time response in this case is shown in Figure 6.24.

(ii) For a unit ramp input,

$$u(z) = \frac{Tz}{(z - 1)^2}$$

and the system output response (with $T = 1$) is given by

$$y(z) = \frac{0.63z}{(z - 1)^2(z - 0.37)} = \frac{0.63z}{z^3 - 2.37z^2 + 1.74z - 0.37}.$$

Using the long division method, we obtain the first few output samples as

$$y(z) = 0.63z^{-2} + 1.5z^{-3} + 2.45z^{-4} + 3.43z^{-5} + \ldots$$

and the output response is given as

$$y(nT) = 0.63\delta(t - 2) + 1.5\delta(t - 3) + 2.45\delta(t - 4) + 3.43\delta(t - 5) + \ldots,$$

as shown in Figure 6.25.

Example 6.20

The open-loop block diagram of a system with a zero-order hold is shown in Figure 6.26. Calculate and plot the system response when a step input is applied to the system, assuming that $T = 1$ s.

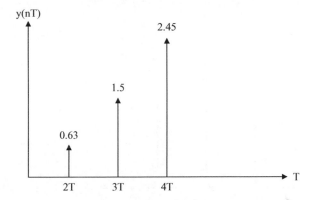

Figure 6.25 Ramp input time response of Example 6.19

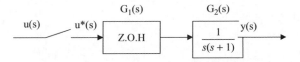

Figure 6.26 Open-loop system with zero-order hold

Solution

The transfer function of the zero-order hold is

$$G_1(s) = \frac{1 - e^{-Ts}}{s}$$

and that of the plant is

$$G(s) = \frac{1}{s(s+1)}.$$

For this system we can write

$$y(s) = u^*(s)G_1G_2(s)$$

and

$$y^*(s) = u^*(s)[G_1G_2]*(s)$$

or, taking z-transforms,

$$y(z) = u(z)G_1G_2(z).$$

Now, $T = 1$ s and

$$G_1G_2(s) = \frac{1 - e^{-s}}{s^2(s+1)}$$

or, by partial fraction expansion,

$$G_1G_2(s) = (1 - e^{-s})\left(\frac{1}{s^2} - \frac{1}{s} + \frac{1}{s+1}\right)$$

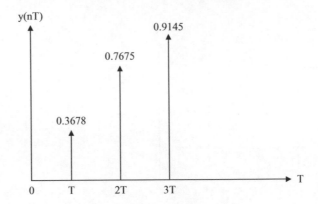

Figure 6.27 Output response

and the z-transform is given by

$$G_1G_2(z) = (1 - z^{-1})Z\left[\frac{1}{s^2} - \frac{1}{s} + \frac{1}{s+1}\right].$$

From the z-transform tables we obtain

$$G_1G_2(z) = (1 - z^{-1})\left[\frac{z}{(z-1)^2} - \frac{z}{z-1} + \frac{z}{z-e^{-1}}\right] = \frac{ze^{-1} + 1 - 2e^{-1}}{(z-1)(z-e^{-1})}$$

$$= \frac{0.3678z + 0.2644}{z^2 - 1.3678z + 0.3678}.$$

After long division we obtain the time response

$$y(nT) = 0.3678\delta(t-1) + 0.7675\delta(t-2) + 0.9145\delta(t-3) + \ldots,$$

shown in Figure 6.27.

6.3.3 Closed-Loop Systems

Some examples of manipulating the closed-loop system block diagrams are given in this section.

Example 6.21

The block diagram of a closed-loop sampled data system is shown in Figure 6.28. Derive an expression for the transfer function of the system.

Solution

For the system in Figure 6.28 we can write

$$e(s) = r(s) - H(s)y(s) \tag{6.38}$$

and

$$y(s) = e^*(s)G(s). \tag{6.39}$$

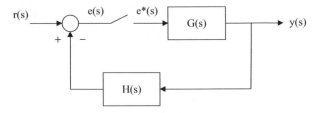

Figure 6.28   Closed-loop sampled data system

Substituting (6.39) into (6.38),

$$e(s) = r(s) - G(s)H(s)e^*(s) \tag{6.40}$$

or

$$e^*(s) = r^*(s) - GH^*(s)e^*(s)$$

and, solving for $e^*(s)$, we obtain

$$e^*(s) = \frac{r^*(s)}{1 + GH^*(s)} \tag{6.41}$$

and, from (6.39),

$$y(s) = G(s)\frac{r^*(s)}{1 + GH^*(s)}. \tag{6.42}$$

The sampled output is then

$$y^*(s) = \frac{r^*(s)G^*(s)}{1 + GH^*(s)} \tag{6.43}$$

Writing (6.43) in z-transform format,

$$y(z) = \frac{r(z)G(z)}{1 + GH(z)} \tag{6.44}$$

and the transfer function is given by

$$\frac{y(z)}{r(z)} = \frac{G(z)}{1 + GH(z)}. \tag{6.45}$$

Example 6.22

The block diagram of a closed-loop sampled data system is shown in Figure 6.29. Derive an expression for the output function of the system.

Solution

For the system in Figure 6.29 we can write

$$y(s) = e(s)G(s) \tag{6.46}$$

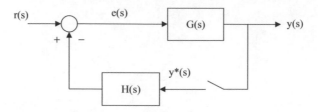

Figure 6.29   Closed-loop sampled data system

and

$$e(s) = r(s) - H(s)y^*(s). \tag{6.47}$$

Substituting (6.47) into (6.46), we obtain

$$y(s) = G(s)r(s) - G(s)H(s)y^*(s) \tag{6.48}$$

or

$$y^*(s) = Gr^*(s) - GH^*(s)y^*(s). \tag{6.49}$$

Solving for $y^*(s)$, we obtain

$$y^*(s) = \frac{Gr^*(s)}{1 + GH^*(s)} \tag{6.50}$$

and

$$y(z) = \frac{Gr(z)}{1 + GH(z)}. \tag{6.51}$$

Example 6.23

The block diagram of a closed-loop sampled data control system is shown in Figure 6.30. Derive an expression for the transfer function of the system.

Solution

The A/D converter can be approximated with an ideal sampler. Similarly, the D/A converter at the output of the digital controller can be approximated with a zero-order hold. Denoting

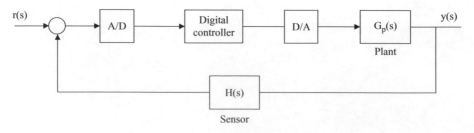

Figure 6.30   Closed-loop sampled data system

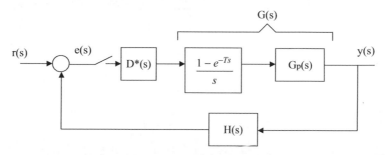

Figure 6.31   Equivalent diagram for Example 6.23

the digital controller by $D(s)$ and combining the zero-order hold and the plant into $G(s)$, the block diagram of the system can be drawn as in Figure 6.31. For this system can write

$$e(s) = r(s) - H(s)y(s) \tag{6.52}$$

and

$$y(s) = e^*(s)D^*(s)G(s). \tag{6.53}$$

Note that the digital computer is represented as $D^*(s)$. Using the above two equations, we can write

$$e(s) = r(s) - D^*(s)G(s)H(s)e^*(s) \tag{6.54}$$

or

$$e^*(s) = r^*(s) - D^*(s)GH^*(s)e^*(s)$$

and, solving for $e^*(s)$, we obtain

$$e^*(s) = \frac{r^*(s)}{1 + D^*(s)GH^*(s)} \tag{6.55}$$

and, from (6.53),

$$y(s) = D^*(s)G(s)\frac{r^*(s)}{1 + D^*(s)GH^*(s)}. \tag{6.56}$$

The sampled output is then

$$y^*(s) = \frac{r^*(s)D^*(s)G^*(s)}{1 + D^*(s)GH^*(s)}, \tag{6.57}$$

Writing (6.57) in z-transform format,

$$y(z) = \frac{r(z)D(z)G(z)}{1 + D(z)GH(z)} \tag{6.58}$$

and the transfer function is given by

$$\frac{y(z)}{r(z)} = \frac{D(z)G(z)}{1 + D(z)GH(z)}. \tag{6.59}$$

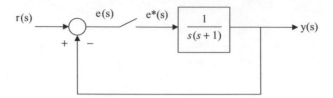

Figure 6.32   Closed-loop system

6.3.4 Closed-Loop Time Response

The closed-loop time response of a sampled data system can be obtained by finding the inverse z-transform of the output function. Some examples are given below.

Example 6.24

A unit step signal is applied to the sampled data digital system shown in Figure 6.32. Calculate and plot the output response of the system. Assume that $T = 1$ s.

Solution

The output response of this system is given in (6.44) as

$$y(z) = \frac{r(z)G(z)}{1 + GH(z)}.$$

where

$$r(z) = \frac{z}{z-1}, \quad G(z) = \frac{z(1-e^{-T})}{(z-1)(z-e^{-T})}, \quad H(z) = 1;$$

thus,

$$y(z) = \frac{z/z - 1}{1 + (z(1-e^{-T})/(z-1)(z-e^{-T}))} \frac{z(1-e^{-T})}{(z-1)(z-e^{-T})}.$$

Simplifying,

$$y(z) = \frac{z^2(1-e^{-T})}{(z^2 - 2ze^{-T} + e^{-T})(z-1)}.$$

Since $T = 1$,

$$y(z) = \frac{0.632z^2}{z^3 - 1.736z^2 + 1.104z - 0.368}.$$

After long division we obtain the first few terms

$$y(z) = 0.632z^{-1} + 1.096z^{-2} + 1.25z^{-3} + \ldots.$$

The first 10 samples of the output response are shown in Figure 6.33.

6.4 EXERCISES

1. A function $y(t) = 2\sin 4t$ is sampled every $T = 0.1$ s. Find the z-transform of the resultant number sequence.

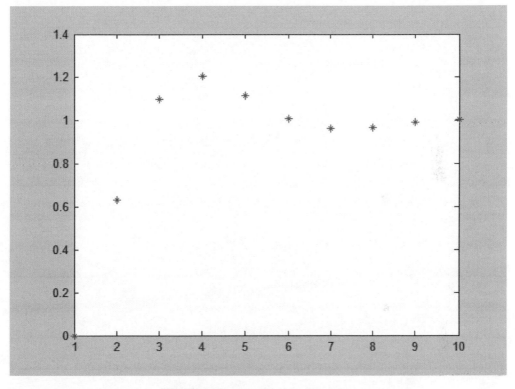

Figure 6.33   First 10 output samples

2. Find the z-transform of the function $y(t) = 3t$.

3. Find the inverse z-transform of the function

$$y(z) = \frac{z}{(z+1)(z-1)}.$$

4. The output response of a system is described with the z-transform

$$y(z) = \frac{z}{(z+0.5)(z-0.2)}.$$

   (i) Apply the final value theorem to calculate the final value of the output when a unit step input is applied to the system.

   (ii) Check your results by finding the inverse z-transform of $y(z)$.

5. Find the inverse z-transform of the following functions using both long division and the method of partial fractions. Compare the two methods.

   (i) $y(z) = \dfrac{0.2z}{(z-1)(z-0.5)}$

   (ii) $y(z) = \dfrac{0.1(z+1)}{(z-0.2)(z-1)}$

   (iii) $y(z) = \dfrac{0.2}{(z-3)(z-1)}$

   (iv) $y(z) = \dfrac{z(z-1)}{(z-2)^2}$

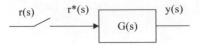

Figure 6.34   Open-loop system for Exercise 6

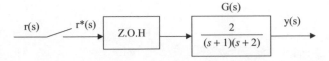

Figure 6.35   Open-loop system with zero-order hold for Exercise 10

6. Consider the open-loop system given in Figure 6.34. Find the output response when a unit step is applied, if

$$G(s) = \frac{0.2}{s(s+1)}.$$

7. Draw the output waveform of Exercise 6.

8. Find the z-transform of the following function, assuming that $T = 0.5$ s:

$$y(s) = \frac{s+1}{(s-1)(s+3)}.$$

9. Find the z-transforms of the following functions, using z-transform tables:

(i) $y(s) = \dfrac{s+1}{s(s+2)}$ (ii) $y(s) = \dfrac{s}{(s+1)^2}$

(iii) $y(s) = \dfrac{s^2}{(s+1)^2(s+2)}$ (iv) $y(s) = \dfrac{0.4}{s(s+1)(s+2)}$

10. Figure 6.35 shows an open-loop system with a zero-order hold. Find the output response when a unit step input is applied. Assume that $T = 0.1$ s and

$$G(s) = \frac{2}{(s+1)(s+2)}.$$

11. Repeat *Exercise 10* for the case where the plant transfer function is given by

(i) $G(s) = \dfrac{0.1}{s(s+2)}$ (ii) $G(s) = \dfrac{2s}{(s+1)(s+4)}$

12. Derive an expression for the transfer function of the closed-loop system whose block diagram is shown in Figure 6.36.

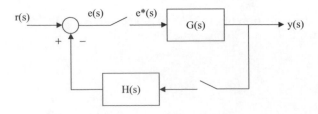

Figure 6.36   Closed-loop system for Exercise 12

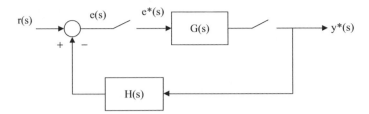

Figure 6.37 Closed-loop system for Exercise 13

13. Derive an expression for the output function of the closed-loop system whose block diagram is shown in Figure 6.37.

FURTHER READING

[Astrom and Wittenmark, 1984] Astrom, K.J. and Wittenmark, B. Computer Controlled Systems. Prentice Hall, Englewood Cliffs, NJ, 1984.

[D'Azzo and Houpis, 1966] D'Azzo, J.J. and Houpis, C.H. Feedback Control System Analysis and Synthesis, 2nd edn., McGraw-Hill, New York, 1966.

[Dorf, 1992] Dorf, R.C. Modern Control Systems, 6th edn., Addison-Wesley, Reading, MA, 1992.

[Evans, 1954] Evans, W.R. Control System Dynamics. McGraw-Hill, New York, 1954.

[Houpis and Lamont, 1962] Houpis, C.H. and Lamont, G.B. Digital Control Systems: Theory, Hardware, Software, 2nd edn., McGraw-Hill, New York, 1962.

[Hsu and Meyer, 1968] Hsu, J.C. and Meyer, A.U. Modern Control Principles and Applications. McGraw-Hill, New York, 1968.

[Jury, 1958] Jury, E.I. Sampled-Data Control Systems. John Wiley & Sons, Inc., New York, 1958.

[Katz, 1981] Katz, P. Digital Control Using Microprocessors. Prentice Hall, Englewood Cliffs, NJ, 1981.

[Kuo, 1963] Kuo, B.C. Analysis and Synthesis of Sampled-Data Control Systems. Englewood Cliffs, NJ, Prentice Hall, 1963.

[Lindorff, 1965] Lindorff, D.P. Theory of Sampled-Data Control Systems. John Wiley & Sons, Inc., New York, 1965.

[Ogata, 1990] Ogata, K. Modern Control Engineering, 2nd edn., Prentice Hall, Englewood Cliffs, NJ, 1990.

[Phillips and Harbor, 1988] Phillips, C.L. and Harbor, R.D. Feedback Control Systems. Englewood Cliffs, NJ, Prentice Hall, 1988.

[Raven, 1995] Raven, F.H. Automatic Control Engineering, 5th edn., McGraw-Hill, New York, 1995.

[Smith, 1972] Smith, C.L. Digital Computer Process Control. Intext Educational Publishers, Scranton, PA, 1972.

[Strum and Kirk, 1988] Strum, R.D. and Kirk, D.E. First Principles of Discrete Systems and Digital Signal Processing. Addison-Wesley, Reading, MA, 1988.

[Tou, 1959] Tou, J. Digital and Sampled-Data Control Systems. McGraw-Hill, New York, 1959.

7

System Time Response Characteristics

In this chapter we investigate the time response of a sampled data system and compare it with the response of a similar continuous system. In addition, the mapping between the s-domain and the z-domain is examined, the important time response characteristics of continuous systems are revised and their equivalents in the discrete domain are discussed.

7.1 TIME RESPONSE COMPARISON

An example closed-loop discrete-time system with a zero-order hold is shown in Figure 7.1(a). The continuous-time equivalent of this system is also shown in Figure 7.1(b), where the sampler (A/D converter) and the zero-order hold (D/A converter) have been removed. We shall now derive equations for the step responses of both systems and then plot and compare them.

As described in Chapter 6, the transfer function of the above discrete-time system is given by

$$\frac{y(z)}{r(z)} = \frac{G(z)}{1 + G(z)},\qquad(7.1)$$

where

$$r(z) = \frac{z}{z-1}\qquad(7.2)$$

and the z-transform of the plant is given by

$$G(s) = \frac{1 - e^{-sT}}{s^2(s+1)}.$$

Expanding by means of partial fractions, we obtain

$$G(s) = (1 - e^{-sT})\left(\frac{1}{s^2} - \frac{1}{s} + \frac{1}{s+1}\right)$$

Microcontroller Based Applied Digital Control   D. Ibrahim
© 2006 John Wiley & Sons, Ltd

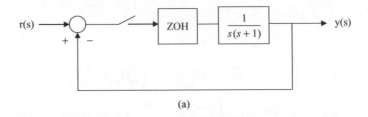

(a)

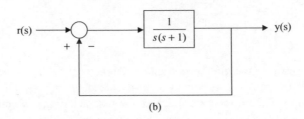

(b)

Figure 7.1   (a) Discrete system and (b) its continuous-time equivalent

and the z-transform is

$$G(z) = (1 - z^{-1})Z \left\{ \frac{1}{s^2} - \frac{1}{s} + \frac{1}{s+1} \right\}.$$

From z-transform tables we obtain

$$G(z) = (1 - z^{-1}) \left[\frac{Tz}{(z-1)^2} - \frac{z}{z-1} + \frac{z}{z - e^{-T}} \right].$$

Setting $T = 1$s and simplifying gives

$$G(z) = \frac{0.368z + 0.264}{z^2 - 1.368z + 0.368}.$$

Substituting into (7.1), we obtain the transfer function

$$\frac{y(z)}{r(z)} = \frac{G(z)}{1 + G(z)} = \frac{0.368z + 0.264}{z^2 - z + 0.632},$$

and then using (7.2) gives the output

$$y(z) = \frac{z(0.368z + 0.264)}{(z-1)(z^2 - z + 0.632)}.$$

The inverse z-transform can be found by long division: the first several terms are

$$y(z) = 0.368z^{-1} + z^{-2} + 1.4z^{-3} + 1.4z^{-4} + 1.15z^{-5} + 0.9z^{-6} + 0.8z^{-7} + 0.87z^{-8}$$
$$+ 0.99z^{-9} + \dots$$

and the time response is given by

$$y(nT) = 0.368\delta(t-1) + \delta(t-2) + 1.4\delta(t-3) + 1.4\delta(t-4) + 1.15\delta(t-5)$$
$$+ 0.9\delta(t-6) + 0.8\delta(t-7) + 0.87\delta(t-8) + \dots .$$

From Figure 7.1(b), the equivalent continuous-time system transfer function is

$$\frac{y(s)}{r(s)} = \frac{G(s)}{1 + G(s)} = \frac{1/(s(s+1))}{1 + (1/s(s+1))} = \frac{1}{s^2 + s + 1}.$$

Since $r(s) = 1/s$, the output becomes

$$y(s) = \frac{1}{s(s^2 + s + 1)}.$$

To find the inverse Laplace transform we can write

$$y(s) = \frac{1}{s} - \frac{s+1}{s^2+s+1} = \frac{1}{s} - \frac{s+0.5}{(s+0.5)^2 - 0.5^2} - \frac{0.5}{(s+0.5)^2 - 0.5^2}.$$

From inverse Laplace transform tables we find that the time response is

$$y(t) = 1 - e^{-0.5t}\left(\cos 0.5t + 0.577 \sin 0.5t\right).$$

Figure 7.2 shows the time responses of both the discrete-time system and its continuous-time equivalent. The response of the discrete-time system is accurate only at the sampling instants. As shown in the figure, the sampling process has a destabilizing effect on the system.

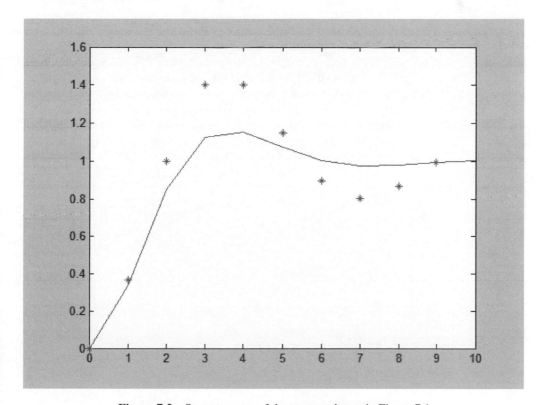

Figure 7.2 Step response of the system shown in Figure 7.1

7.2 TIME DOMAIN SPECIFICATIONS

The performance of a control system is usually measured in terms of its response to a step input. The step input is used because it is easy to generate and gives the system a nonzero steady-state condition, which can be measured.

Most commonly used time domain performance measures refer to a second-order system with the transfer function:

$$\frac{y(s)}{r(s)} = \frac{\omega_n^2}{s^2 + 2\zeta\omega_n s + \omega_n^2},$$

where ω_n is the undamped natural frequency of the system and ζ is the damping ratio of the system.

When a second-order system is excited with a unit step input, the typical output response is as shown in Figure 7.3. Based on this figure, the following performance parameters are usually defined: maximum overshoot; peak time; rise time; settling time; and steady-state error.

The maximum overshoot, M_p, is the peak value of the response curve measured from unity. This parameter is usually quoted as a percentage. The amount of overshoot depends on the damping ratio and directly indicates the relative stability of the system.

The peak time, T_p, is defined as the time required for the response to reach the first peak of the overshoot. The system is more responsive when the peak time is smaller, but this gives rise to a higher overshoot.

The rise time, T_r, is the time required for the response to go from 0 % to 100 % of its final value. It is a measure of the responsiveness of a system, and smaller rise times make the system more responsive.

The settling time, T_s, is the time required for the response curve to reach and stay within a range about the final value. A value of 2–5 % is usually used in performance specifications.

The steady-state error, E_{ss}, is the error between the system response and the reference input value (unity) when the system reaches its steady-state value. A small steady-tate error is a requirement in most control systems. In some control systems, such as position control, it is one of the requirements to have no steady-state error.

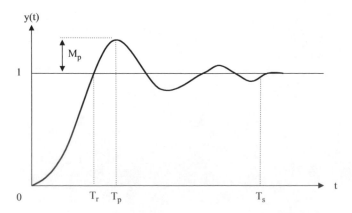

Figure 7.3   Second-order system unit step response

Having introduced the parameters, we are now in a position to give formulae for them (readers who are interested in the derivation of these formulae should refer to books on control theory). The maximum overshoot occurs at at peak time ($t = T_p$) and is given by

$$M_p = e^{-(\zeta\pi/\sqrt{1-\zeta^2})},$$

i.e. overshoot is directly related to the system damping ratio – the lower the damping ratio, the higher the overshoot. Figure 7.4 shows the variation of the overshoot (expressed as a percentage) with the damping ratio.

The peak time is obtained by differentiating the output response with respect to time, letting this equal zero. It is given by

$$T_p = \frac{\pi}{\omega_d},$$

where

$$\omega_d = \omega_n^2\sqrt{1 - \zeta^2}$$

is the damped natural frequency.

The rise time is obtained by setting the output response to 1 and finding the time. It is given by

$$T_r = \frac{\pi - \beta}{\omega_d},$$

where

$$\beta = \tan^{-1}\frac{w_d}{\zeta\omega_n}.$$

The settling time is usually specified for a 2 % or 5 % tolerance band, and is given by

$$T_s = \frac{4}{\zeta\omega_n} \quad \text{(for 2\% settling time),}$$

$$T_s = \frac{3}{\zeta\omega_n} \quad \text{(for 5\% settling time).}$$

The steady-state error can be found by using the final value theorem, i.e. if the Laplace transform of the output response is $y(s)$, then the final value (steady-state value) is given by

$$\lim_{s\to 0} sy(s),$$

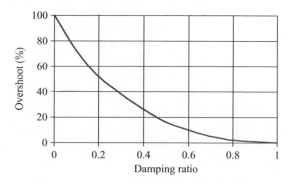

Figure 7.4 Variation of overshoot with damping ratio

and the steady-state error when a unit step input is applied can be found from

$$E_{ss} = 1 - \lim_{s \to 0} s\, y(s).$$

Example 7.1

Determine the performance parameters of the system given in Section 7.1 with closed-loop transfer function

$$\frac{y(s)}{r(s)} = \frac{1}{s^2 + s + 1}.$$

Solution

Comparing this system with the standard second-order system transfer function

$$\frac{y(s)}{r(s)} = \frac{\omega_n^2}{s^2 + 2\zeta\omega_n s + \omega_n^2},$$

we find that $\zeta = 0.5$ and $\omega_n = 1$ rad/s. Thus, the damped natural frequency is

$$\omega_d = \omega_n^2 \sqrt{1 - \zeta^2} = 0.866 \text{rad/s}.$$

The peak overshoot is

$$M_p = e^{-(\zeta\pi/\sqrt{1-\zeta^2})} = 0.16$$

or 16 %. The peak time is

$$T_p = \frac{\pi}{\omega_d} = 3.627 \text{ s}$$

The rise time is

$$T_r = \frac{\pi - \beta}{\omega_n};$$

since

$$\beta = \tan^{-1} \frac{\omega_d}{\zeta\omega_n} = 1.047,$$

we have

$$T_r = \frac{\pi - \beta}{\omega_n} = \frac{\pi - 1.047}{1} = 2.094 \text{ s}$$

The settling time (2 %) is

$$T_s = \frac{4}{\zeta\omega_n} = 8 \text{ s},$$

and the settling time (5 %) is

$$T_s = \frac{3}{\zeta\omega_n} = 6 \text{ s}.$$

Finally, the steady state error is

$$E_{ss} = 1 - \lim_{s \to 0} s\, y(s) = 1 - \lim_{s \to 0} s \frac{1}{s(s^2 + s + 1)} = 0.$$

7.3 MAPPING THE s-PLANE INTO THE z-PLANE

The pole locations of a closed-loop continuous-time system in the *s*-plane determine the behaviour and stability of the system, and we can shape the response of a system by positioning its poles in the *s*-plane. It is desirable to do the same for the sampled data systems. This section describes the relationship between the *s*-plane and the *z*-plane and analyses the behaviour of a system when the closed-loop poles are placed in the *z*-plane.

First of all, consider the mapping of the left-hand side of the *s*-plane into the *z*-plane. Let $s = \sigma + j\omega$ describe a point in the *s*-plane. Then, along the $j\omega$ axis,

$$z = e^{sT} = e^{\sigma T} e^{j\omega T}.$$

But $\sigma = 0$ so we have

$$z = e^{j\omega T} = \cos \omega T + j \sin \omega T = 1 \angle \omega T.$$

Hence, the pole locations on the imaginary axis in the *s*-plane are mapped onto the unit circle in the *z*-plane. As ω changes along the imaginary axis in the *s*-plane, the angle of the poles on the unit circle in the *z*-plane changes.

If ω is kept constant and σ is increased in the left-hand *s*-plane, the pole locations in the *z*-plane move towards the origin, away from the unit circle. Similarly, if σ is decreased in the left-hand *s*-plane, the pole locations in the *z*-plane move away from the origin in the *z*-plane. Hence, the entire left-hand *s*-plane is mapped into the interior of the unit circle in the *z*-plane. Similarly, the right-hand *s*-plane is mapped into the exterior of the unit circle in the *z*-plane. As far as the system stability is concerned, a sampled data system will be stable if the closed-loop poles (or the zeros of the characteristic equation) lie within the unit circle. Figure 7.5 shows the mapping of the left-hand *s*-plane into the *z*-plane.

As shown in Figure 7.6, lines of constant σ in the *s*-plane are mapped into circles in the *z*-plane with radius $e^{\sigma T}$. If the line is on the left-hand side of the *s*-plane then the radius of the circle in the *z*-plane is less than 1. If on the other hand the line is on the right-hand side of the *s*-plane then the radius of the circle in the *z*-plane is greater than 1. Figure 7.7 shows the corresponding pole locations between the *s*-plane and the *z*-plane.

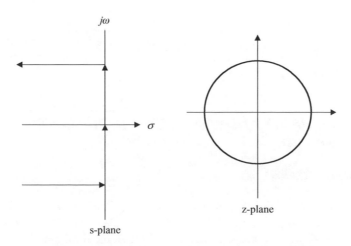

Figure 7.5 Mapping the left-hand *s*-plane into the *z*-plane

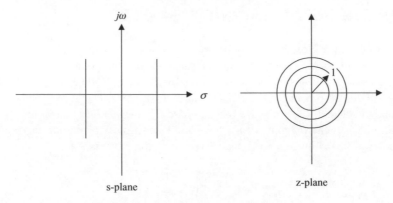

Figure 7.6   Mapping the lines of constant σ

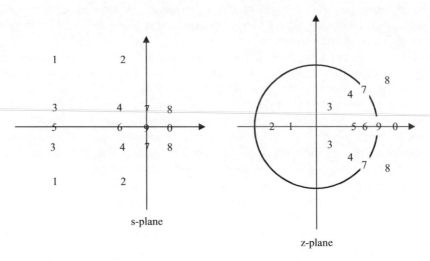

Figure 7.7   Poles in the s-plane and their corresponding z-plane locations

The time responses of a sampled data system based on its pole positions in the z-plane are shown in Figure 7.8. It is clear from this figure that the system is stable if all the closed-loop poles are within the unit circle.

7.4   DAMPING RATIO AND UNDAMPED NATURAL FREQUENCY IN THE z-PLANE

7.4.1   Damping Ratio

As shown in Figure 7.9(a), lines of constant damping ratio in the s-plane are lines where $\zeta = \cos \alpha$ for a given damping ratio. The locus in the z-plane can then be obtained by the substitution $z = e^{sT}$. Remembering that we are working in the third and fourth quadrants in

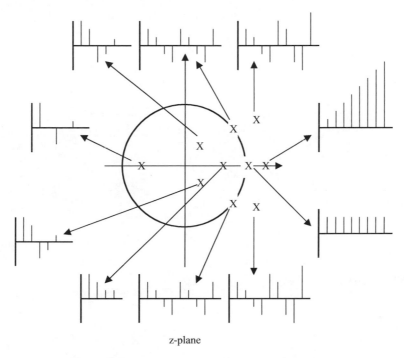

z-plane

Figure 7.8   Time response of z-plane pole locations

the s-plane where s is negative, we get

$$z = e^{-\sigma\omega T} e^{j\omega T}.$$ (7.3)

Since, from Figure 7.9(a),

$$\sigma = \tan\left(\frac{\pi}{2} - \cos^{-1}\zeta\right),$$ (7.4)

substituting in (7.3) we have

$$z = \exp\left[-\omega T \tan\left(\frac{\pi}{2} - \cos^{-1}\zeta\right)\right] e^{j\omega T}.$$ (7.5)

Equation (7.5) describes a logarithmic spiral in the z-plane as shown in Figure 7.9(b). The spiral starts from $z = 1$ when $\omega = 0$. Figure 7.10 shows the lines of constant damping ratio in the z-plane for various values of ζ.

7.4.2 Undamped Natural Frequency

As shown in Figure 7.11, the locus of constant undamped natural frequency in the s-plane is a circle with radius ω_n. From this figure, we can write

$$\omega^2 + \sigma^2 = \omega_n^2 \quad \text{or} \quad \sigma = \sqrt{\omega_n^2 - \omega^2}.$$ (7.6)

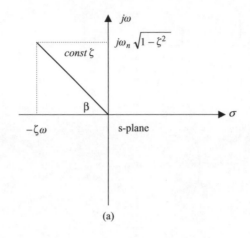

(a)

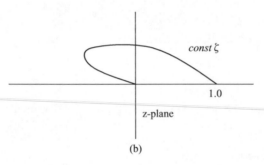

(b)

Figure 7.9 (a) Line of constant damping ratio in the s-plane, and (b) the corresponding locus in the z-plane

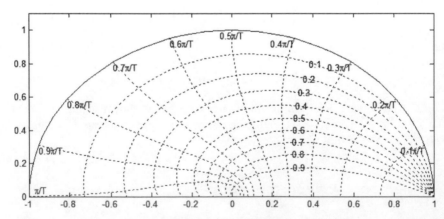

Figure 7.10 Lines of constant damping ratio for different ζ. The vertical lines are the lines of constant ω_n

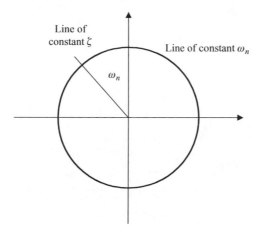

Figure 7.11   Locus of constant ω_n in the s-plane

Thus, remembering that s is negative, we have

$$z = e^{-sT} = e^{-\sigma T} e^{-j\omega T} = \exp\left[-T(\sqrt{\omega_n^2 - \omega^2})\right] e^{-j\omega T} \qquad (7.7)$$

The locus of constant ω_n in the z-plane is given by (7.7) and is shown in Figure 7.10 as the vertical lines. Notice that the curves are given for values of ω_n ranging from $\omega_n = \pi/10T$ to $\omega_n = \pi/T$.

Notice that the loci of constant damping ratio and the loci of undamped natural frequency are usually shown on the same graph.

7.5 DAMPING RATIO AND UNDAMPED NATURAL FREQUENCY USING FORMULAE

In Section 7.4 above we saw how to find the damping ratio and the undamped natural frequency of a system using a graphical technique. Here, we will derive equations for calculating the damping ratio and the undamped natural frequency.

The damping ratio and the natural frequency of a system in the z-plane can be determined if we first of all consider a second-order system in the s-plane:

$$G(s) = \frac{\omega_n^2}{s^2 + 2\zeta\omega_n s + \omega_n^2}. \qquad (7.8)$$

The poles of this system are at

$$s_{1,2} = -\zeta\omega_n \pm j\omega_n\sqrt{1 - \zeta^2}. \qquad (7.9)$$

We can now find the equivalent z-plane poles by making the substitution $z = e^{sT}$, i.e.

$$z = e^{sT} = e^{-\zeta\omega_n T} \angle \pm \omega_n T\sqrt{1 - \zeta^2}, \qquad (7.10)$$

which we can write as

$$z = r \angle \pm \theta, \tag{7.11}$$

where

$$r = e^{-\zeta \omega_n T} \quad \text{or} \quad \zeta \omega_n T = -\ln r \tag{7.12}$$

and

$$\theta = \omega_n T \sqrt{1 - \zeta^2}. \tag{7.13}$$

From (7.12) and (7.13) we obtain

$$\frac{\zeta}{\sqrt{1 - \zeta^2}} = \frac{-\ln r}{\theta}$$

or

$$\zeta = \frac{-\ln r}{\sqrt{(\ln r)^2 + \theta^2}}, \tag{7.14}$$

and from (7.12) and (7.14) we obtain

$$\omega_n = \frac{1}{T} \sqrt{(\ln r)^2 + \theta^2}. \tag{7.15}$$

Example 7.2

Consider the system described in Section 7.1 with closed-loop transfer function

$$\frac{y(z)}{r(z)} = \frac{G(z)}{1 + G(z)} = \frac{0.368z + 0.264}{z^2 - z + 0.632}.$$

Find the damping ratio and the undamped natural frequency. Assume that $T = 1$ s.

Solution

We need to find the poles of the closed-loop transfer function. The system characteristic equation is $1 + G(z) = 0$,
i.e.

$$z^2 - z + 0.632 = (z - 0.5 - j0.618)(z - 0.5 + j0.618) = 0,$$

which can be written in polar form as

$$z_{1,2} = 0.5 \pm j0.618 = 0.795 \angle \pm 0.890 = r \angle \pm \theta$$

(see (7.11)). The damping ratio is then calculated using (7.14) as

$$\zeta = \frac{-\ln r}{\sqrt{(\ln r)^2 + \theta^2}} = \frac{-\ln 0.795}{\sqrt{(\ln 0.795)^2 + 0.890^2}} = 0.25,$$

and from (7.15) the undamped natural frequency is, taking $T = 1$,

$$\omega_n = \frac{1}{T} \sqrt{(\ln r)^2 + \theta^2} = \sqrt{(\ln 0.795)^2 + 0.890^2} = 0.92.$$

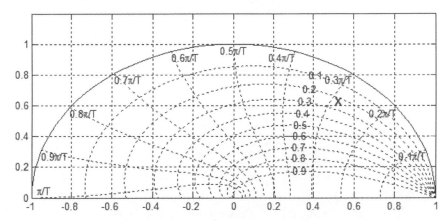

Figure 7.12   Finding ζ and ω_n graphically

Example 7.3

Find the damping ratio and the undamped natural frequency for Example 7.2 using the graphical method.

Solution

The characteristic equation of the system is found to be

$$z^2 - z + 0.632 = (z - 0.5 - j0.618)(z - 0.5 + j0.618) = 0$$

and the poles of the closed-loop system are at

$$z_{1,2} = 0.5 \pm j0.618.$$

Figure 7.12 shows the loci of the constant damping ratio and the loci of the undamped natural frequency with the poles of the closed-loop system marked with an $\times$ on the graph. From the graph we can read the damping ratio as 0.25 and the undamped natural frequency as

$$\omega_n = \frac{0.29\pi}{T} = 0.91.$$

7.6 EXERCISES

1. Find the damping ratio and the undamped natural frequency of the sampled data systems whose characteristic equations are given below
   (a) $z^2 - z + 2 = 0$
   (b) $z^2 - 1 = 0$
   (c) $z^2 - z + 1 = 0$
   (d) $z^2 - 0.81 = 0$

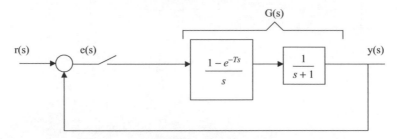

Figure 7.13 System for Exercise 2

2. Consider the closed-loop system of Figure 7.13. Assume that $T = 1$ s.
   (a) Calculate the transfer function of the system.
   (b) Calculate and plot the unit step response at the sampling instants.
   (c) Calculate the damping factor and the undamped natural frequency of the system.

3. Consider the closed-loop system of Figure 7.13. Do not assume a value for T.
   (a) Calculate the transfer function of the system.
   (b) Calculate the damping factor and the undamped natural frequency of the system.
   (c) What will be the steady state error if a unit step input is applied?

4. A unit step input is applied to the system in Figure 7.13. Calculate:
   (a) the percentage overshoot;
   (b) the peak time;
   (c) the rise time;
   (d) settling time to 5 %.

5. The closed-loop transfer functions of four sampled data systems are given below. Calculate the percentage overshoots and peak times.
   (a) $G(z) = \dfrac{1}{z^2 + z + 2}$
   (b) $G(z) = \dfrac{1}{z^2 + 2z + 1}$
   (c) $G(z) = \dfrac{1}{z^2 - z + 1}$
   (d) $G(z) = \dfrac{\frac{1}{2}}{z^2 + z + 4}$

6. The s-plane poles of a continuous-time system are at $s = -1$ and $s = -2$. Assuming $T = 1$ s, calculate the pole locations in the z-plane.

7. The s-plane poles of a continuous-time system are at $s_{1,2} = -0.5 \pm j0.9$. Assuming $T = 1$ s, calculate the pole locations in the z-plane. Calculate the damping ratio and the undamped natural frequency of the system using a graphical technique.

FURTHER READING

[D'Azzo and Houpis, 1966] D'Azzo, J.J. and Houpis, C.H. Feedback Control System Analysis and Synthesis, 2nd edn., McGraw-Hill, New York, 1966.

[Dorf, 1992]   Dorf, R.C. Modern Control Systems, 6th edn. Addison-Wesley, Reading, MA, 1992.

[Evans, 1954]   Evans, W.R. Control System Dynamics, McGraw-Hill, New York, 1954.

[Houpis and Lamont, 1962] Houpis, C.H. and Lamont, G.B. Digital Control Systems: Theory, Hardware, Software, 2nd edn., McGraw-Hill, New York, 1962.

[Hsu and Meyer, 1968] Hsu, J.C. and Meyer, A.U. Modern Control Principles and Applications. McGraw-Hill, New York, 1968.

[Jury, 1958]   Jury, E.I. Sampled-Data Control Systems. John Wiley & Sons, Inc., New York, 1958.

[Katz, 1981]   Katz, P. Digital Control Using Microprocessors. Prentice Hall, Englewood Cliffs, NJ, 1981.

[Kuo, 1963]   Kuo, B.C. Analysis and Synthesis of Sampled-Data Control Systems. Prentice Hall, Englewood Cliffs, NJ, 1963.

[Lindorff, 1965]   Lindorff, D.P. Theory of Sampled-Data Control Systems. John Wiley & Sons, Inc., New York, 1965.

[Ogata, 1990]   Ogata, K. Modern Control Engineering, 2nd edn., Prentice Hall, Englewood Cliffs, NJ, 1990.

[Phillips and Harbor, 1988] Phillips, C.L. and Harbor R.D. Feedback Control Systems. Englewood Cliffs, NJ, Prentice Hall, 1988.

[Raven, 1995]   Raven, F.H. Automatic Control Engineering, 5th edn., McGraw-Hill, New York, 1995.

[Strum and Kirk, 1988] Strum, R.D. and Kirk D.E. First Principles of Discrete Systems and Digital Signal Processing. Addison-Wesley, Reading, MA, 1988.

8

System Stability

This chapter is concerned with the various techniques available for the analysis of the stability of discrete-time systems.

Suppose we have a closed-loop system transfer function

$$\frac{Y(z)}{R(z)} = \frac{G(z)}{1 + GH(z)} = \frac{N(z)}{D(z)},$$

where $1 + GH(z) = 0$ is also known as the characteristic equation. The stability of the system depends on the location of the poles of the closed-loop transfer function, or the roots of the characteristic equation $D(z) = 0$. It was shown in Chapter 7 that the left-hand side of the s-plane, where a continuous system is stable, maps into the interior of the unit circle in the z-plane. Thus, we can say that a system in the z-plane will be stable if all the roots of the characteristic equation, $D(z) = 0$, lie inside the unit circle.

There are several methods available to check for the stability of a discrete-time system:

- Factorize $D(z) = 0$ and find the positions of its roots, and hence the position of the closed-loop poles.

- Determine the system stability without finding the poles of the closed-loop system, such as Jury's test.

- Transform the problem into the s-plane and analyse the system stability using the well-established s-plane techniques, such as frequency response analysis or the Routh–Hurwitz criterion.

- Use the root-locus graphical technique in the z-plane to determine the positions of the system poles.

The various techniques described in this section will be illustrated with examples.

8.1 FACTORIZING THE CHARACTERISTIC EQUATION

The stability of a system can be determined if the characteristic equation can be factorized. This method has the disadvantage that it is not usually easy to factorize the characteristic equation. Also, this type of test can only tell us whether or not a system is stable as it is. It does not tell us about the margin of stability or how the stability is affected if the gain or some other parameter is changed in the system.

Microcontroller Based Applied Digital Control   D. Ibrahim
© 2006 John Wiley & Sons, Ltd

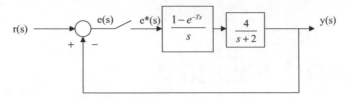

Figure 8.1   Closed-loop system

Example 8.1

The block diagram of a closed-loop system is shown in Figure 8.1. Determine whether or not the system is stable. Assume that $T = 1$ s.

Solution

The closed-loop system transfer function is

$$\frac{Y(z)}{R(z)} = \frac{G(z)}{1 + G(z)}, \tag{8.1}$$

where

$$G(z) = Z\left\{\left[\frac{1 - e^{-Ts}}{s}\frac{4}{s+2}\right]\right\} = (1 - z^{-1})Z\left\{\left[\frac{4}{s(s+2)}\right]\right\} = (1 - z^{-1})\frac{2z(1 - e^{-2T})}{(z - 1)(z - e^{-2T})}$$

$$= \frac{2(1 - e^{-2T})}{z - e^{-2T}}. \tag{8.2}$$

For $T = 1$ s,

$$G(z) = \frac{1.729}{z - 0.135}.$$

The roots of the characteristic equation are $1 + G(z) = 0$, or $1 + 1.729/(z - 0.135) = 0$, the solution of which is $z = -1.594$ which is outside the unit circle, i.e. the system is not stable.

Example 8.2

For the system given in Example 8.1, find the value of T for which the system is stable.

Solution

From (8.2),

$$G(z) = \frac{2(1 - e^{-2T})}{z - e^{-2T}}.$$

The roots of the characteristic equation are $1 + G(z) = 0$, or $1 + 2(1 - e^{-2T})/(z - e^{-2T}) = 0$, giving

$$z - e^{-2T} + 2(1 - e^{-2T}) = 0$$

or

$$z = 3e^{-2T} - 2.$$

The system will be stable if the absolute value of the root is inside the unit circle, i.e.

$$|3e^{-2T} - 2| < 1,$$

from which we get

$$2T < \ln\left(\tfrac{1}{3}\right) \quad \text{or} \quad T < 0.549.$$

Thus, the system will be stable as long as the sampling time $T < 0.549$.

8.2 JURY'S STABILITY TEST

Jury's stability test is similar to the Routh–Hurwitz stability criterion used for continuous-time systems. Although Jury's test can be applied to characteristic equations of any order, its complexity increases for high-order systems.

To describe Jury's test, express the characteristic equation of a discrete-time system of order n as

$$F(z) = a_n z^n + a_{n-1} z^{n-1} + \ldots + a_1 z + a_0 = 0, \tag{8.3}$$

where $a_n > 0$. We now form the array shown in Table 8.1. The elements of this array are defined as follows:

- The elements of each of the even-numbered rows are the elements of the preceding row, in reverse order.

- The elements of the odd-numbered rows are defined as:

$$b_k = \begin{vmatrix} a_0 & a_{n-k} \\ a_n & a_k \end{vmatrix}, \quad c_k = \begin{vmatrix} b_0 & b_{n-k-1} \\ n_{n-1} & b_k \end{vmatrix}, \quad d_k = \begin{vmatrix} c_0 & c_{n-2-k} \\ c_{n-2} & c_k \end{vmatrix}, \quad \ldots.$$

Table 8.1 Array for Jury's stability tests

z^0	z^1	z^2	$\ldots$	z^{n-k}	$\ldots$	z^{n-1}	z^n
a_0	a_1	a_2	$\ldots$	a_{n-k}	$\ldots$	a_{n-1}	a_n
a_n	a_{n-1}	a_{n-2}	$\ldots$	a_k	$\ldots$	a_1	a_0
b_0	b_1	b_2	$\ldots$	b_{n-k}	$\ldots$	b_{n-1}	
b_{n-1}	b_{n-2}	b_{n-3}	$\ldots$	b_{k-1}	$\ldots$	b_0	
c_0	c_1	c_2	$\ldots$	c_{n-k}	$\ldots$		
c_{n-2}	c_{n-3}	c_{n-4}	$\ldots$	c_{k-2}	$\ldots$		
$\ldots$	$\ldots$	$\ldots$	$\ldots$	$\ldots$			
$\ldots$	$\ldots$	$\ldots$	$\ldots$	$\ldots$			
l_0	l_1	l_2	l_3				
l_3	l_2	l_1	l_0				
m_0	m_1	m_2					

The necessary and sufficient conditions for the characteristic equation (8.3) to have roots inside the unit circle are given as

$$F(1) > 0, \quad (-1)^n F(-1) > 0, \quad |a_0| < a_n, \tag{8.4}$$

$$\begin{aligned} |b_0| &> b_{n-1} \\ |c_0| &> c_{n-2} \\ |d_0| &> d_{n-3} \\ &\cdots \\ &\cdots \\ |m_0| &> m_2. \end{aligned} \tag{8.5}$$

Jury's test is then applied as follows:

• Check the three conditions given in (8.4) and stop if any of these conditions is not satisfied.

• Construct the array given in Table 8.1 and check the conditions given in (8.5). Stop if any condition is not satisfied.

Jury's test can become complex as the order of the system increases. For systems of order 2 and 3 the test reduces to the following simple rules. Given the second-order system characteristic equation

$$F(z) = a_2 z^2 + a_1 z + a_0 = 0, \quad \text{where } a_2 > 0,$$

no roots of the system characteristic equation will be on or outside the unit circle provided that

$$F(1) > 0, \quad F(-1) > 0, \quad |a_0| < a_2..$$

Given the third-order system characteristic equation

$$F(z) = a_3 z^3 + a_2 z^2 + a_1 z + a_0 = 0, \quad \text{where } a_3 > 0,$$

no roots of the system characteristic equation will be on or outside the unit circle provided that

$$F(1) > 0, \quad F(-1) < 0, \quad |a_0| < a_3,$$

$$\left| \det \begin{bmatrix} a_0 & a_3 \\ a_3 & a_0 \end{bmatrix} \right| > \left| \det \begin{bmatrix} a_0 & a_1 \\ a_3 & a_2 \end{bmatrix} \right|.$$

Examples are given below.

Example 8.3

The closed-loop transfer function of a system is given by

$$\frac{G(z)}{1 + G(z)},$$

where

$$G(z) = \frac{0.2z + 0.5}{z^2 - 1.2z + 0.2}.$$

Determine the stability of this system using Jury's test.

Solution

The characteristic equation is

$$1 + G(z) = 1 + \frac{0.2z + 0.5}{z^2 - 1.2z + 0.2} = 0$$

or

$$z^2 - z + 0.7 = 0.$$

Applying Jury's test,

$$F(1) = 0.7 > 0, \quad F(-1) = 2.7 > 0, \quad 0.7 < 1.$$

All the conditions are satisfied and the system is stable.

Example 8.4

The characteristic equation of a system is given by

$$1 + G(z) = 1 + \frac{K(0.2z + 0.5)}{z^2 - 1.2z + 0.2} = 0.$$

Determine the value of K for which the system is stable.

Solution

The characteristic equation is

$$z^2 + z(0.2K - 1.2) + 0.5K = 0, \quad \text{where } K > 0.$$

Applying Jurys's test,

$$F(1) = 0.7K - 0.2 > 0, \quad F(-1) = 0.3K + 2.2 > 0, \quad 0.5K < 1.$$

Thus, the system is stable for $0.285 < K < 2$.

Example 8.5

The characteristic equation of a system is given by

$$F(z) = z^3 - 2z^2 + 1.4z - 0.1 = 0.$$

Determine the stability of the system.

Solution

Applying Jury's test, $a_3 = 1$, $a_2 = -2$, $a_1 = 1.4$, $a_0 = -0.1$ and

$$F(1) = 0.3 > 0, \quad F(-1) = -4.5 < 0, \quad 0.1 < 1.$$

The first conditions are satisfied. Applying the other condition,

$$\left\| \begin{bmatrix} -0.1 & 1 \\ 1 & -0.1 \end{bmatrix} \right\| = -0.99 \quad \text{and} \quad \left\| \begin{bmatrix} -0.1 & 1.4 \\ 1 & -2 \end{bmatrix} \right\| = -1.2;$$

since $|0.99| < |-1.2|$, the system is not stable.

8.3   ROUTH–HURWITZ CRITERION

The stability of a sampled data system can be analysed by transforming the system characteristic equation into the s-plane and then applying the well-known Routh–Hurwitz criterion.

A bilinear transformation is usually used to transform the left-hand s-plane into the interior of the unit circle in the z-plane. For this transformation, z is replaced by

$$z = \frac{1 + w}{1 - w}. \tag{8.6}$$

Given the characteristic equation in w,

$$F(w) = b_n w^n + b_{n-1} w^{n-1} + \ldots + b_1 w + b_0 = 0,$$

then the Routh–Hurwitz array is formed as follows:

$$
\begin{array}{c|cccc}
w^n & b_n & b_{n-2} & b_{n-4} & \cdots \\
w^{n-1} & b_{n-1} & b_{n-3} & b_{n-5} & \cdots \\
w^{n-2} & c_1 & c_2 & c_3 & \cdots \\
\cdots & \cdots & \cdots & \cdots & \cdots \\
w^1 & j_1 & & & \\
w^0 & k_1 & & &
\end{array}
$$

The first two rows are obtained from the equation directly and the other rows are calculated as follows:

$$c_1 = \frac{b_{n-1} b_{n-2} - b_n b_{n-3}}{b_{n-1}},$$

$$c_2 = \frac{b_{n-1} b_{n-4} - b_n b_{n-5}}{b_{n-1}},$$

$$c_3 = \frac{b_{n-1} b_{n-6} - b_n b_{n-7}}{b_{n-1}},$$

$$d_1 = \frac{c_1 b_{n-3} - b_{n-1} c_2}{c_1},$$

$$\ldots\ldots$$

The Routh–Hurwitz criterion states that the number of roots of the characteristic equation in the right hand s-plane is equal to the number of sign changes of the coefficients in the first column of the array. Thus, for a stable system all coefficients in the first column must have the same sign.

Example 8.6

The characteristic equation of a sampled data system is given by

$$z^2 - z + 0.7 = 0.$$

Determine the stability of the system using the Routh–Hurwitz criterion.

Solution

Transforming the characteristic equation into the w-plane gives

$$\left(\frac{1 + w}{1 - w} \right)^2 - \frac{1 + w}{1 - w} + 0.7 = 0,$$

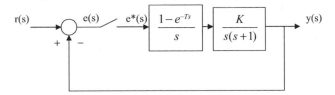

Figure 8.2   Closed-loop system

or

$$2.7w^2 + 0.6w + 0.7 = 0.$$

Forming the Routh–Hurwitz array,

$$
\begin{array}{c|cc}
w^2 & 2.7 & 0.7 \\
w^1 & 0.6 & 0 \\
w^0 & 0.7 &
\end{array}
$$

there are no sign changes in the first column and thus the system is stable.

Example 8.7

The block diagram of a sampled data system is shown in Figure 8.2. Use the Routh–Hurwitz criterion to determine the value of K for which the system is stable. Assume that $K > 0$ and $T = 1$ s.

Solution

The characteristic equation is $1 + G(z) = 0$, where

$$G(s) = \frac{1 - e^{-Ts}}{s} \frac{K}{s(s+1)}.$$

The z-transform is given by

$$G(z) = (1 - z^{-1})Z\left\{ \frac{K}{s^2(s+1)} \right\},$$

which gives

$$G(z) = \frac{K(0.368z + 0.264)}{(z-1)(z-0.368)}.$$

The characteristic equation is

$$1 + \frac{K(0.368z + 0.264)}{(z-1)(z-0.368)} = 0,$$

or

$$z^2 - z(1.368 - 0.368K) + 0.368 + 0.264K = 0.$$

Transforming into the w-plane gives

$$\left(\frac{1+w}{1-w}\right)^2 - \left(\frac{1+w}{1-w}\right)(1.368 - 0.368K) + 0.368 + 0.264K = 0$$

or

$$w^2(2.736 - 0.104K) + w(1.264 - 0.528K) + 0.632K = 0.$$

We can now form the Routh–Hurwitz array

$$
\begin{array}{c|cc}
w^2 & 2.736 - 0.104K & 0.632K \\
w^1 & 1.264 - 0.528K & 0 \\
w^0 & 0.632K &
\end{array}
$$

The system is stable if there is no sign change in the first column. Thus, for stability,

$$1.264 - 0.528K > 0$$

or

$$K < 2.4.$$

8.4   ROOT LOCUS

The root locus is one of the most powerful techniques used to analyse the stability of a closed-loop system. This technique is also used to design controllers with required time response characteristics. The root locus is a plot of the locus of the roots of the characteristic equation as the gain of the system is varied. The rules of the root locus for discrete-time systems are identical to those for continuous systems. This is because the roots of an equation $Q(z) = 0$ in the z-plane are the same as the roots of $Q(s) = 0$ in the s-plane. Even though the rules are the same, the interpretation of the root locus is quite different in the s-plane and the z-plane. For example, a continuous system is stable if the roots are in the left-hand s-plane. A discrete-time system, on the other hand, is stable if the roots are inside the unit circle. The construction and the rules of the root locus for continuous-time systems are described in many textbooks. In this section only the important rules for the construction of the discrete-time root locus are given, with worked examples.

Given the closed-loop system transfer function

$$\frac{G(z)}{1 + GH(z)},$$

we can write the characteristic equation as $1 + kF(z) = 0$, and the root locus can then be plotted as k is varied. The rules for constructing the root locus can be summarized as follows:

1. The locus starts on the poles of $F(z)$ and terminate on the zeros of $F(z)$.

2. The root locus is symmetrical about the real axis.

3. The root locus includes all points on the real axis to the left of an odd number of poles and zeros.

4. If $F(z)$ has zeros at infinity, the root locus will have asymptotes as $k \to \infty$. The number of asymptotes is equal to the number of poles n_p, minus the number of zeros n_z. The angles of the asymptotes are given by

$$\theta = \frac{180r}{n_p - n_z}, \quad \text{where } r = \pm 1, \pm 3, \pm 5, \dots.$$

The asymptotes intersect the real axis at σ, where

$$\sigma = \frac{\sum \text{poles of } F(z) - \sum \text{zeros of } F(z)}{n_p - n_z}.$$

5. The breakaway points on the real axis of the root locus are at the roots of

$$\frac{dF(z)}{dz} = 0.$$

6. If a point is on the root locus, the value of k is given by

$$1 + kF(z) = 0 \quad \text{or} \quad k = -\frac{1}{F(z)}.$$

Example 8.8

A closed-loop system has the characteristic equation

$$1 + GH(z) = 1 + K\frac{0.368(z + 0.717)}{(z - 1)(z - 0.368)} = 0.$$

Draw the root locus and hence determine the stability of the system.

Solution

Applying the rules:

1. The above equation is in the form $1 + kF(z) = 0$, where

$$F(z) = \frac{0.368(z + 0.717)}{(z - 1)(z - 0.368)}.$$

The system has two poles at $z = 1$ and at $z = 0.368$. There are two zeros, one at $z = -0.717$ and the other at minus infinity. The locus will start at the two poles and terminate at the two zeros.

2. The section on the real axis between $z = 0.368$ and $z = 1$ is on the locus. Similarly, the section on the real axis between $z = -\infty$ and $z = -0.717$ is on the locus.

3. Since $n_p - n_z = 1$, there is one asymptote and the angle of this asymptote is

$$\theta = \frac{180r}{n_p - n_z} = \pm 180° \quad \text{for } r = \pm 1.$$

Note that since the angles of the asymptotes are $\pm 180°$ it is meaningless to find the real axis intersection point of the asymptotes.

4. The breakaway points can be found from

$$\frac{dF(z)}{dz} = 0,$$

or

$$0.368(z - 1)(z - 0.368) - 0.368(z + 0.717)(2z - 1.368) = 0,$$

which gives

$$z^2 + 1.434z - 1.348 = 0$$

and the roots are at

$$z = -2.08 \text{ and } z = 0.648.$$

5. The value of k at the breakaway points can be calculated from

$$k = -\frac{1}{F(z)}\Big|_{z=-2.08, 0.648}$$

which gives $k = 15$ and $k = 0.196$.

The root locus of the system is shown in Figure 8.3. The locus is a circle starting from the poles, breaking away at $z = 0.648$ on the real axis, and then joining the real axis at $z = -2.08$.

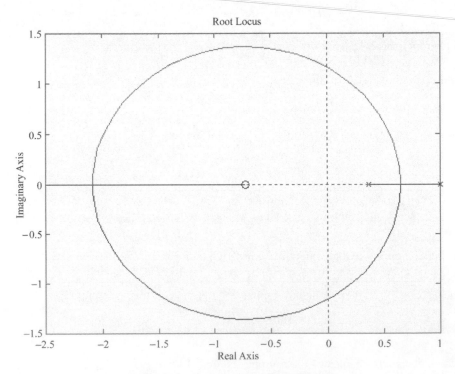

Figure 8.3   Root locus for Example 8.8

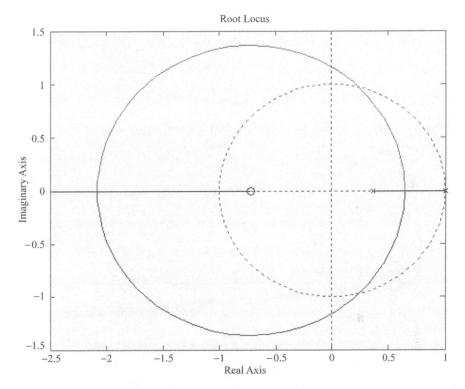

Figure 8.4   Root locus with unit circle

At this point one part of the locus moves towards the zero at $z = -0.717$ and the other moves towards the zero at $-\infty$.

Figure 8.4 shows the root locus with the unit circle drawn on the same axis. The system will become marginally stable when the locus is on the unit circle. The value of k at these points can be found either from Jury's test or by using the Routh–Hurwitz criterion.

Using Jury's test, the characteristic equation is

$$1 + K\frac{0.368(z + 0.717)}{(z - 1)(z - 0.368)} = 0,$$

or

$$z^2 - z(1.368 - 0.368K) + 0.368 + 0.263K = 0.$$

Applying Jury's test

$$F(1) = 0.631 \quad \text{for } K > 0.$$

Also,

$$|0.263K + 0.368| < 1$$

which gives $K = 2.39$ for marginal stability of the system.

Example 8.9

For Example 8.8, calculate the value of k for which the damping factor is $\zeta = 0.7$.

Solution

In Figure 8.5 the root locus of the system is redrawn with the lines of constant damping factor and constant natural frequency.

From the figure, the roots when $\zeta = 0.7$ are read as $s_{1,2} = 0.61 \pm j0.25$ (see Figure 8.6). The value of k can now be calculated as

$$k = -\left. \frac{1}{F(z)} \right|_{z=0.61 \pm j0.25}$$

which gives $k = 0.324$.

Example 8.10

A closed-loop system has the characteristic equation

$$1 + GH(z) = 1 + K \frac{(z - 0.2)}{z^2 - 1.5z + 0.5} = 0.$$

Draw the root locus and hence determine the stability of the system. What will be the value of K for a damping factor $\zeta > 0.6$ and a natural frequency of $\omega_n > 0.6\,\text{rad/s}$?

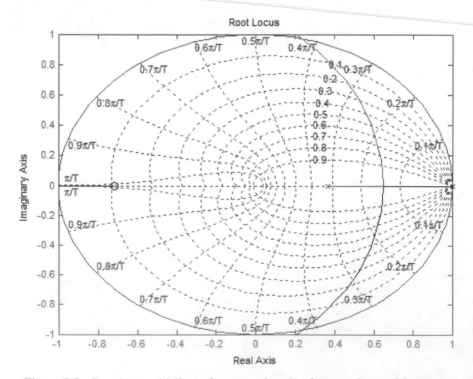

Figure 8.5   Root locus with lines of constant damping factor and natural frequency

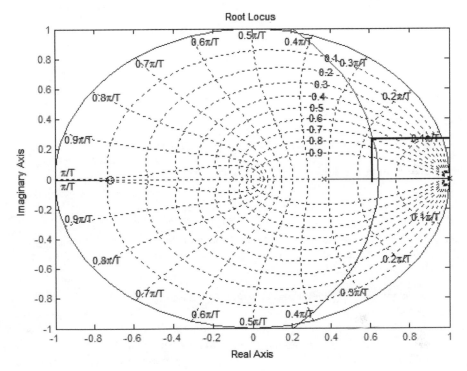

Figure 8.6   Reading the roots when $\zeta = 0.7$

Solution

The above equation is in the form $1 + kF(z) = 0$, where

$$F(z) = \frac{z - 0.2}{z^2 - 1.5z + 0.5}.$$

The system has two poles at $z = 1$ and at $z = 0.5$. There are two zeros, one at $z = -0.2$ and the other at infinity. The locus will start from the two poles and terminate at the two zeros.

1. The section on the real axis between $z = 0.5$ and $z = 1$ is on the locus. Similarly, the section on the real axis between $z = -\infty$ and $z = 0.2$ is on the locus.

2. Since $n_p - n_z = 1$, there is one asymptote and the angle of this asymptote is

$$\theta = \frac{180r}{n_p - n_z} = \pm 180° \quad \text{for } r = \pm 1$$

Note that since the angle of the asymptotes are $\pm 180°$ it meaningless to find the real axis intersection point of the asymptotes.

3. The breakaway points can be found from

$$\frac{dF(z)}{dz} = 0$$

or

$$(z^2 - 1.5z + 0.5) - (z - 0.2)(2z - 1.5) = 0,$$

which gives

$$z^2 - 0.4z - 0.2 = 0$$

and the roots are at

$$z = -0.290 \quad \text{and} \quad z = 0.689.$$

4. The value of k at the breakaway points can be calculated from

$$k = -\frac{1}{F(z)}\bigg|_{z=-0.290, 0.689}$$

which gives $k = 0.12$ and $k = 2.08$. The root locus of the system is shown in Figure 8.7. It is clear from this plot that the system is always stable since all poles are inside the unit circle for all values of k.

Lines of constant damping factor and constant angular frequency are plotted on the same axis in Figure 8.8.

Assuming that $T = 1$ s, $\omega_n > 0.6$ if the roots are on the left-hand side of the constant angular frequency line $\omega_n = 0.2\pi/T$. The damping factor will be greater than 0.6 if the roots are below the constant damping ratio line $\zeta = 0.6$. A point satisfying these properties has been chosen

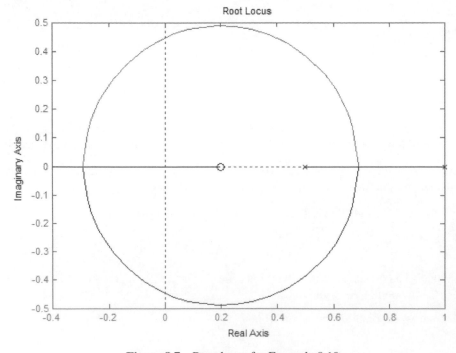

Figure 8.7 Root locus for Example 8.10

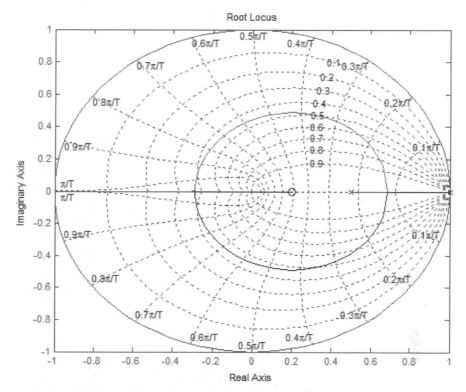

Figure 8.8 Root locus with lines of constant damping factor and natural frequency

and shown in Figure 8.9. The roots at this point are given as $s_{1,2} = 0.55 \pm j0.32$. The value of k can now be calculated as

$$k = -\frac{1}{F(z)}\bigg|_{z=0.55 \pm j0.32}$$

which gives $k = 0.377$.

8.5 NYQUIST CRITERION

The Nyquist criterion is one of the widely used stability analysis techniques in the s-plane, based on the frequency response of the system. To determine the frequency response of a continuous system transfer function $G(s)$, we replace s by $j\omega$ and use the transfer function $G(j\omega)$. In the s-plane, the Nyquist criterion is based on the plot of the magnitude $|GH(j\omega)|$ against the angle $\angle GH(j\omega)$ as ω is varied.

In a similar manner, the frequency response of a transfer function $G(z)$ in the z-plane can be obtained by making the substitution $z = e^{j\omega T}$. The Nyquist plot in the z-plane can then be obtained by plotting the magnitude of $|GH(z)|_{z=e^{j\omega T}}$ against the angle $\angle GH(z)|_{z=e^{j\omega T}}$ as ω is varied. The criterion is then

$$Z = N + P,$$

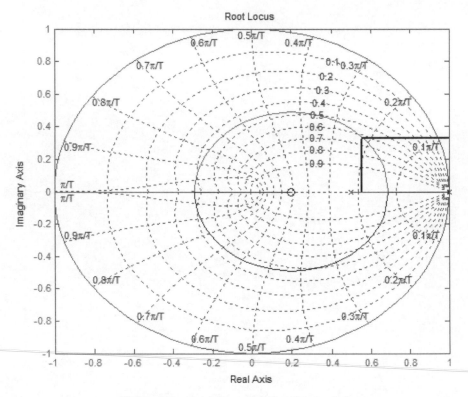

Figure 8.9   Point for $\zeta > 0.6$ and $\omega_n > 0.6$

where N is the number of clockwise circles around the point -1, P the number of poles of $GH(z)$ that are outside the unit circle, and Z the number of zeros of $GH(z)$ that are outside the unit circle.

For a stable system, Z must be equal to zero, and hence the number of anticlockwise circles around the point -1 must be equal to the number of poles of $GH(z)$.

If $GH(z)$ has no poles outside the unit circle then the criterion becomes simple and for stability the Nyquist plot must not encircle the point -1.

An example is given below.

Example 8.11

The transfer function of a closed-loop sampled data system is given by

$$\frac{G(z)}{1 + GH(z)},$$

where

$$GH(z) = \frac{0.4}{(z - 0.5)(z - 0.2)}.$$

Determine the stability of this system using the Nyquist criterion. Assume that $T = 1$ s.

Solution

Setting $z = e^{j\omega T} = \cos \omega T + j \sin \omega T = \cos \omega + j \sin \omega$,

$$G(z)|_{z=e^{j\omega T}} = \frac{0.4}{(\cos \omega + j \sin \omega - 0.5)(\cos \omega + j \sin \omega - 0.2)}$$

or

$$G(z)|_{z=e^{j\omega T}} = \frac{0.4}{(\cos^2 \omega - \sin^2 \omega - 0.7 \cos \omega + 0.1) + j(2 \sin \omega \cos \omega - 0.7 \sin \omega)}.$$

This has magnitude

$$|G(z)| = \frac{0.4}{\sqrt{(\cos^2 \omega - \sin^2 \omega - 0.7 \cos \omega + 0.1)^2 + (2 \sin \omega \cos \omega - 0.7 \sin \omega)^2}}$$

and phase

$$\angle G(z) = \tan^{-1} \frac{2 \sin \omega \cos \omega - 0.7 \sin \omega}{\cos^2 \omega - \sin^2 \omega - 0.7 \cos \omega + 0.1}.$$

Table 8.2 lists the variation of the magnitude of $G(z)$ with the phase angle. The Nyquist plot for this example is shown in Figure 8.10. Since $N = 0$ and $P = 0$, the closed-loop system has no poles outside the unit circle in the x-plane and the system is stable.

The Nyquist diagram can also be plotted by transforming the system into the w-plane and then using the standard s-plane Nyquist criterion. An example is given below.

Example 8.12

The open-loop transfer function of a unity feedback sampled data system is given by

$$G(z) = \frac{z}{(z - 1)(z - 0.4)}.$$

Derive expressions for the magnitude and the phase of $|G(z)|$ by transforming the system into the w-plane.

Solution

The w transformation is defined as

$$z = \frac{1 + w}{1 - w}$$

which gives

$$G(w) = \frac{(1 + w)/(1 - w)}{((1 + w/1 - w) - 1)((1 + w/1 - w) - 0.4)} = \frac{1 + w}{2w(0.6 + 1.4w)},$$

or

$$G(w) = \frac{1 + w}{1.2w + 2.8w^2}.$$

Table 8.2 Magnitude and phase of $G(z)$

w	$|G(z)|$	$\angle G(z)$
0	1.0000E+000	0
1.0472E−001	9.8753E−001	−1.9430E+001
2.0944E−001	9.5248E−001	−3.8460E+001
3.1416E−001	9.0081E−001	−5.6779E+001
4.1888E−001	8.3961E−001	−7.4209E+001
5.2360E−001	7.7510E−001	−9.0690E+001
6.2832E−001	7.1166E−001	−1.0625E+002
7.3304E−001	6.5193E−001	−1.2096E+002
8.3776E−001	5.9719E−001	−1.3492E+002
9.4248E−001	5.4789E−001	−1.4820E+002
1.0472E+000	5.0395E−001	−1.6089E+002
1.1519E+000	4.6505E−001	−1.7308E+002
1.2566E+000	4.3075E−001	1.7518E+002
1.3614E+000	4.0058E−001	1.6384E+002
1.4661E+000	3.7408E−001	1.5283E+002
1.5708E+000	3.5082E−001	1.4213E+002
1.6755E+000	3.3044E−001	1.3168E+002
1.7802E+000	3.1259E−001	1.2147E+002
1.8850E+000	2.9698E−001	1.1146E+002
1.9897E+000	2.8337E−001	1.0162E+002
2.0944E+000	2.7154E−001	9.1945E+001
2.1991E+000	2.6130E−001	8.2401E+001
2.3038E+000	2.5250E−001	7.2974E+001
2.4086E+000	2.4501E−001	6.3646E+001
2.5133E+000	2.3872E−001	5.4404E+001
2.6180E+000	2.3354E−001	4.5232E+001
2.7227E+000	2.2939E−001	3.6118E+001
2.8274E+000	2.2622E−001	2.7050E+001
2.9322E+000	2.2399E−001	1.8015E+001
3.0369E+000	2.2266E−001	9.0018E+000

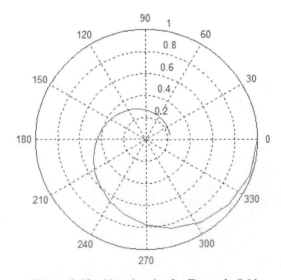

Figure 8.10 Nyquist plot for Example 8.11

But since the w-plane can be regarded an analogue of the s-plane, i.e. $s = \sigma + jw$ in the s-plane and $w = \sigma_w + jw_w$ in the z-plane, for the frequency response we can set

$$w = jw_w$$

which yields

$$G(jw_w) = \frac{1 + w_w}{1.2jw_w - 2.8w^2_w}.$$

The magnitude and the phase are then given by

$$|G(jw_w)| = \frac{1 + w_w}{\sqrt{(1.2w_w)^2 + \left(2.8w^2_w\right)^2}}$$

and

$$\angle G(jw_w) = \tan^{-1} \frac{1.2}{2.8w_w},$$

where w_w is related to w by the expression

$$w_w = \tan\left(\frac{wT}{2}\right).$$

8.6 BODE DIAGRAMS

The Bode diagrams used in the analysis of continuous-time systems are not very practical when used directly in the z-plane. This is because of the $e^{j\omega T}$ term present in the sampled data system transfer functions when the frequency response is to be obtained. However, it is possible to draw the Bode diagrams of sampled data systems by transforming the system into the w-plane by making the substitution

$$z = \frac{1 + w}{1 - w}, \tag{8.7}$$

where the frequency in the w-plane (w_w) is related to the frequency in the s-plane (w) by the expression

$$w_w = \tan\left(\frac{wT}{2}\right). \tag{8.8}$$

It is common in practice to use a similar transformation to the one given above, known as the w'-plane transformation, which gives a closer analogy between the frequency in the s-plane and the w'-plane. The w'-plane transformation defined as

$$w' = \frac{2}{T}\frac{z - 1}{z + 1}, \tag{8.9}$$

or

$$z = \frac{1 + (T/2)w'}{1 - (T/2)w'}, \tag{8.10}$$

and the frequencies in the two planes are related by the expression

$$w' = \frac{2}{T}\tan\frac{wT}{2}. \tag{8.11}$$

Note that for small values of the real frequency (s-plane frequency) such that wT is small, (8.11) reduces to

$$w' = \frac{2}{T}\tan\frac{wT}{2} \approx \frac{2}{T}\left(\frac{wT}{2}\right) = w \qquad (8.12)$$

Thus, the w'-plane frequency is approximately equal to the s-plane frequency. This approximation is only valid for small values of wT such that $\tan(wT/2) \approx wT$, i.e.

$$\frac{wT}{2} \leq \frac{\pi}{10},$$

which can also be written as

$$w \leq \frac{2\pi}{10T}$$

or

$$w \leq \frac{w_s}{10}, \qquad (8.13)$$

where w_s is the sampling frequency in radians per second. The interpretation of this is that the w'-plane and the s-plane frequencies will be approximately equal when the frequency is less than one-tenth of the sampling frequency.

We can use the transformations given in (8.10) and (8.11) to transform a sampled data system into the w'-plane and then use the standard continuous system Bode diagram analysis.

Some example Bode plots for sampled data systems are given below.

Example 8.13

Consider the closed-loop sampled data system given in Figure 8.11. Draw the Bode diagram and determine the stability of this system. Assume that $T = 0.1$ s.

Solution

From Figure 8.11,

$$G(z) = Z\left\{\frac{1 - e^{-sT}}{s}\frac{5}{s+5}\right\} = \frac{1 - e^{-0.5}}{z - e^{-0.5}},$$

or

$$G(z) = \frac{0.393}{z - 0.606}.$$

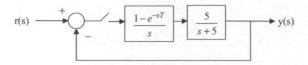

Figure 8.11   Closed-loop system

Transforming the system into the w'-plane gives

$$G(w') = \frac{0.393 - 0.0196w}{0.08w - 0.393},$$

or

$$G(w') = \frac{4.9(1 - 0.05w')}{w' + 4.9}.$$

The magnitude of the frequency response and the phase can now be calculated if we set $w' = jv$ where v is the analogue of true frequency ω. Thus,

$$G(jv) = \frac{4.9(1 - 0.05jv)}{jv + 4.9}.$$

The magnitude is

$$|G(jv)| = \frac{4.9\sqrt{1 + (0.25v)^2}}{\sqrt{v^2 + 4.9^2}}$$

and

$$\angle G(jv) = -\tan^{-1}(0.05) - \tan^{-1}\left(\frac{v}{4.9}\right).$$

The Bode diagram of the system is shown in Figure 8.12. The system is stable.

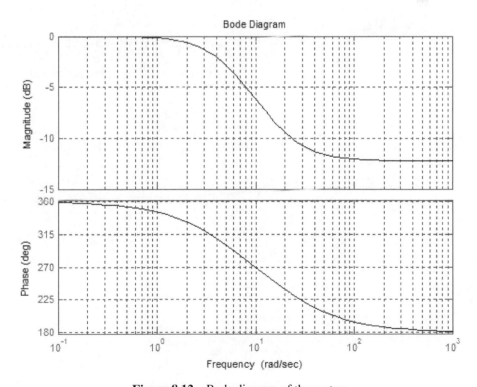

Figure 8.12   Bode diagram of the system

Example 8.14

The loop transfer function of a unity feedback sampled data system is given by

$$G(z) = \frac{0.368z + 0.264}{z^2 - 1.368z + 0.368}.$$

Draw the Bode diagram and analyse the stability of the system. Assume that $T = 1$ s.

Solution

Using the transformation

$$z = \frac{1 + (T/2)w}{1 - (T/2)w} = \frac{1 + 0.5w}{1 - 0.5w},$$

we get

$$G(w) = \frac{0.368(1 + 0.5w/1 - 0.5w) + 0.264}{(1 + 0.5w/1 - 0.5w)^2 - 1.368(1 + 0.5w/1 - 0.5w) + 0.368}$$

or

$$G(w) = -\frac{0.0381(w - 2)(w + 12.14)}{w(w + 0.924)}.$$

To obtain the frequency response, we can replace w with jv, giving

$$G(jv) = -\frac{0.0381(jv - 2)(jv + 12.14)}{jv(jv + 0.924)}.$$

The magnitude is then

$$|G(jv)| = \frac{0.0381\sqrt{v^2 + 2^2}\sqrt{v^2 + 12.14^2}}{v\sqrt{v^2 + 0.924^2}}$$

and

$$\angle G(jv) = \tan^{-1}\frac{v}{2} + \tan^{-1}\frac{v}{12.14} - 90 - \tan^{-1}\frac{v}{0.924}.$$

The Bode diagram is shown in Figure 8.13. The system is stable with a gain margin of 5 dB and a phase margin of 26°.

8.7 EXERCISES

1. Given below are the characteristic equations of some sampled data systems. Using Jury's test, determine if the systems are stable.
   (a) $z^2 - 1.8z + 0.72 = 0$
   (b) $z^2 - 0.5z + 1.2 = 0$
   (c) $z^3 - 2.1z^2 + 2.0z - 0.5 = 0$
   (d) $z^3 - 2.3z^2 + 1.61z - 0.32 = 0$

2. The characteristic equation of a sampled data system is given by

$$(z - 0.5)(z^2 - 0.5z + 1.2) = 0.$$

Determine the stability of the system.

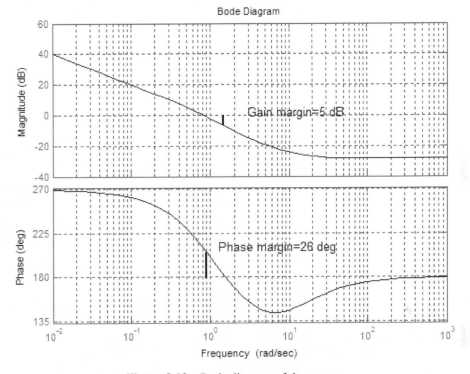

Figure 8.13   Bode diagram of the system

3. For the system shown in Figure 8.14, determine the range of K for stability using Jury's test

4. Repeat Exercise 3 using the Routh–Hurwitz criterion.

5. Repeat Exercise 3 using the root locus.

6. The forward gain of a unity feedback sampled data system is given by

$$G(z) = \frac{K(z - 0.2)}{(z - 0.8)(z - 0.6)}.$$

(a) Write an expression for the closed-loop transfer function of the system.

(b) Draw the root locus of the system and hence determine the stability.

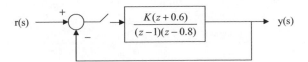

Figure 8.14   System for Exercise 3

7. Check the stability of the transfer function

$$G(z) = \frac{1}{z^3 + 2.7z^2 + 1.5z + 0.2}$$

using:
(a) Jury's test;
(b) the Routh–Hurwitz criterion;
(c) the oot locus

8. Repeat Exercise 7 using the Bode diagram.

9. A process with the transfer function

$$\frac{K}{s(1 + as)}$$

is preceded by a zero-order hold and is connected to form a unity feedback sampled data system.
(a) Assuming the sampling time is T, derive an expression for the closed-loop transfer function of the system.
(b) Draw the root locus of the system and hence determine the value of K for which the system becomes marginally stable.

10. The open-loop transfer function of a sampled data system is given by

$$G(z) = \frac{1}{z^3 - 3.1z^2 + 3.1z - 1.1}.$$

The closed-loop system is formed by using a unity gain feedback. Use Jury's criterion to determine the stability of the system.

11. Use the Bode diagram to determine the stability of the sampled data system given by

$$G(z) = \frac{z}{(z - 1)(z - 0.6)}.$$

12. Repeat Exercise 11 using the Nyquist criterion.

13. The open-loop transfer function of a sampled data system is given by

$$G(z) = \frac{K(z - 0.6)}{(z - 0.8)(z - 0.4)}.$$

(a) Plot the Bode diagram by calculating the frequency response, assuming $K = 1$.
(b) From the Bode diagram determine the phase margin and the gain margin.
(c) Find the value of K for marginal stability.
(d) If the system is marginally stable, determine the frequency of oscillation.

14. The block diagram of a closed-loop sampled data system is shown in Figure 8.15. Determine the range of K for stability by:
(a) finding the roots of the characteristic equation;
(b) using Jury's test;
(c) using the Routh–Hurwitz criterion;

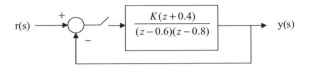

Figure 8.15   Closed-loop system for Exercise 14

(d) using the root locus;
(e) drawing the Bode diagram;
(f) drawing the Nyquist diagram.
Which method would you prefer in this exercise and why?

15. Explain the mapping between the s-plane and the simple w-plane. How are the frequency points mapped?

FURTHER READING

[Bode, 1945]	Bode, H.W. Network Analysis and Feedback Amplifier Design, Van Nostrand, Princeton, NJ, 1945.
[D'Azzo and Houpis, 1966]	D'Azzo, J.J. and Houpis, C.H. Feedback Control System Analysis and Synthesis, 2nd edn., McGraw-Hill. New York, 1966.
[Dorf, 1992]	Dorf, R.C. Modern Control Systems, 6th edn., Addison-Wesley, Reading, MA, 1992.
[Evans, 1948]	Evans, W.R. Graphical analysis of control systems. Trans. AIEE, 67, 1948, pp. 547–551.
[Evans, 1954]	Evans, W.R. Control System Dynamics. McGraw-Hill. New York, 1954.
[Houpis and Lamont, 1962]	Houpis, C.H. and Lamont, G.B. Digital Control Systems: Theory, Hardware, Software, 2nd edn., McGraw-Hill, New York, 1962.
[Jury, 1958]	Jury, E.I. Sampled-Data Control Systems. John Wiley & Sons, Inc., New York, 1958.
[Katz, 1981]	Katz, P. Digital Control Using Microprocessors. Prentice Hall, Englewood Cliffs, NJ, 1981.
[Kuo, 1962]	Kuo, B.C. Automatic Control Systems. Prentice Hall, Englewood Cliffs, NJ, 1962.
[Kuo, 1963]	Kuo, B.C. Analysis and Synthesis of Sampled-Data Control Systems. Prentice Hall, Englewood Cliffs, NJ, 1963.
[Lindorff, 1965]	Lindorff, D.P. Theory of Sampled-Data Control Systems. John Wiley & Sons, Inc., New York, 1965.
[Phillips and Harbor, 1988]	Phillips, C.L. and Harbor, R.D. Feedback Control Systems. Englewood Cliffs, NJ, Prentice Hall, 1988.
[Soliman and Srinath, 1990]	Soliman, S.S. and Srinath, M.D. Continuous and Discrete Signals and Systems. Prentice Hall, Englewood Cliffs, NJ, 1990.
[Strum and Kirk, 1988]	Strum, R.D. and Kirk, D.E. First Principles of Discrete Systems and Digital Signal Processing. Addison-Wesley, Reading, MA, 1988.

9

Discrete Controller Design

The design of a digital control system begins with an accurate model of the process to be controlled. Then a control algorithm is developed that will give the required system response. The loop is closed by using a digital computer as the controller. The computer implements the control algorithm in order to achieve the required response.

Several methods can be used for the design of a digital controller:

- A system transfer function is modelled and obtained in the s-plane. The transfer function is then transformed into the z-plane and the controller is designed in the z-plane.

- System transfer function is modelled as a digital system and the controller is directly designed in the z-plane.

- The continuous system transfer function is transformed into the w-plane. A suitable controller is then designed in the w-plane using the well-established time response (e.g. root locus) or frequency response (e.g. Bode diagram) techniques. The final design is transformed into the z-plane and the algorithm is implemented on the digital computer.

In this chapter we are mainly interested in the design of a digital controller using the first method, i.e. the controller is designed directly in the z-plane.

The procedure for designing the controller in the z-plane can be outlined as follows:

- Derive the transfer function of the system either by using a mathematical approach or by performing a frequency or a time response analysis.

- Transform the system transfer function into the z-plane.

- Design a suitable digital controller in the z-plane.

- Implement the controller algorithm on a digital computer.

A discrete-time system can be in many different forms, depending on the type of input and the type of sensor used. Figure 9.1 shows a discrete-time system where the reference input is an analog signal, and the process output is also an analog signal. Analog-to-digital converters are then used to convert these signals into digital form so that they can be processed by a digital computer. A zero-order hold at the output of the digital controller approximates a D/A converter which produces an analog signal to drive the plant.

In Figure 9.2 the reference input is a digital signal, which is usually set using a keyboard or can be hard-coded into the controller algorithm. The feedback signal is also digital and the

---

Microcontroller Based Applied Digital Control   D. Ibrahim
© 2006 John Wiley & Sons, Ltd

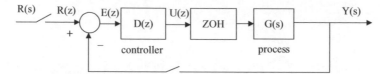

Figure 9.1   Discrete-time system with analog reference input

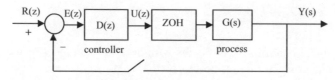

Figure 9.2   Discrete-time system with digital reference input

error signal is formed by the computer after subtracting the feedback signal from the reference input. The digital controller then implements the control algorithm and derives the plant.

9.1   DIGITAL CONTROLLERS

In general, we can make use of the block diagram shown in Figure 9.3 when designing a digital controller. In this figure, $R(z)$ is the reference input, $E(z)$ is the error signal, $U(z)$ is the output of the controller, and $Y(z)$ is the output of the system. $HG(z)$ represents the digitized plant transfer function together with the zero-order hold.

The closed-loop transfer function of the system in Figure 9.3 can be written as

$$\frac{Y(z)}{R(z)} = \frac{D(z)HG(z)}{1 + D(z)HG(z)}. \tag{9.1}$$

Now, suppose that we wish the closed-loop transfer function to be $T(z)$, i.e.

$$T(z) = \frac{Y(z)}{R(z)}. \tag{9.2}$$

Then the required controller that will give this closed-loop response can be found by using (9.1) and (9.2):

$$D(z) = \frac{1}{HG(z)} \frac{T(z)}{1 - T(z)}. \tag{9.3}$$

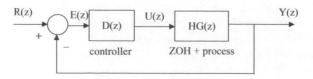

Figure 9.3   Discrete-time system

Equation (9.3) states that the required controller $D(z)$ can be designed if we know the model of the process. The controller $D(z)$ must be chosen so that it is stable and can be realized. One of the restrictions affecting realizability is that $D(z)$ must not have a numerator whose order exceeds that of the numerator. Some common controllers based on (9.3) are described below.

9.1.1 Dead-Beat Controller

The dead-beat controller is one in which a step input is followed by the system but delayed by one or more sampling periods, i.e. the system response is required to be equal to unity at every sampling instant after the application of a unit step input.

The required closed-loop transfer function is then

$$T(z) = z^{-k}, \qquad \text{where } k \geq 1. \tag{9.4}$$

From (9.3), the required digital controller transfer function is

$$D(z) = \frac{1}{HG(z)} \frac{T(z)}{1 - T(z)} = \frac{1}{HG(z)} \left(\frac{z^{-k}}{1 - z^{-k}} \right). \tag{9.5}$$

An example design of a controller using the dead-beat algorithm is given below.

Example 9.1

The open-loop transfer function of a plant is given by

$$G(s) = \frac{e^{-2s}}{1 + 10s}.$$

Design a dead-beat digital controller for the system. Assume that $T = 1$ s.

Solution

The transfer function of the system with a zero-order hold is given by

$$HG(z) = Z\left\{ \frac{1 - e^{-sT}}{s} G(s) \right\} = (1 - z^{-1})Z\left\{ \frac{e^{-2s}}{s(1 + 10s)} \right\}$$

or

$$HG(z) = (1 - z^{-1})z^{-2}Z\left\{ \frac{1}{s(1 + 10s)} \right\} = (1 - z^{-1})z^{-2}Z\left\{ \frac{1/10}{s(s + 1/10)} \right\}.$$

From z-transform tables we obtain

$$HG(z) = (1 - z^{-1})z^{-2} \frac{z(1 - e^{-0.1})}{(z - 1)(z - e^{-0.1})} = z^{-3} \frac{(1 - e^{-0.1})}{1 - e^{-0.1}z^{-1}}$$

or

$$HG(z) = \frac{0.095z^{-3}}{1 - 0.904z^{-1}}.$$

From Equations (9.3) and (9.5),

$$D(z) = \frac{1}{HG(z)} \frac{T(z)}{1 - T(z)} = \frac{1 - 0.904z^{-1}}{0.095z^{-3}} \frac{z^{-k}}{1 - z^{-k}}.$$

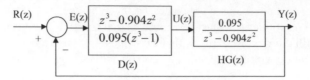

Figure 9.4   Block diagram of the system of Example 9.1

For realizability, we can choose $k \geq 3$. Choosing $k = 3$, we obtain

$$D(z) = \frac{1 - 0.904z^{-1}}{0.095z^{-3}} \frac{z^{-3}}{1 - z^{-3}}$$

or

$$D(z) = \frac{z^3 - 0.904z^2}{0.095(z^3 - 1)}.$$

Figure 9.4 shows the system block diagram with the controller, while Figure 9.5 shows the step response of the system. The output response is unity after 3 s (third sample) and stays at this value. It is important to realize that the response is correct only at the sampling instants and the response can have an oscillatory behaviour between the sampling instants.

The control signal applied to the plant is shown in Figure 9.6. Although the dead-beat controller has provided an excellent response, the magnitude of the control signal may not be acceptable, and it may even saturate in practice.

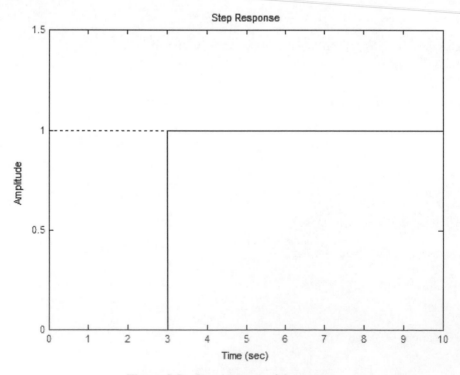

Figure 9.5   Step response of the system

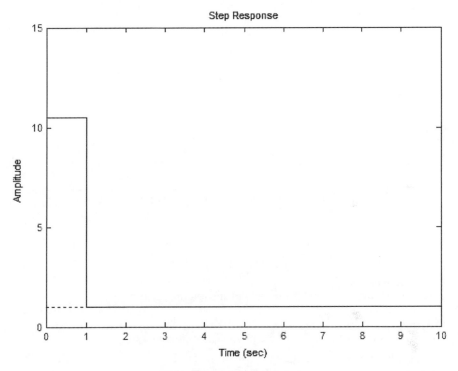

Figure 9.6 Control signal

The dead-beat controller is very sensitive to plant characteristics and a small change in the plant may lead to ringing or oscillatory response.

9.1.2 Dahlin Controller

The Dahlin controller is a modification of the dead-beat controller and produces an exponential response which is smoother than that of the dead-beat controller.

The required response of the system in the s-plane can be shown to be

$$Y(s) = \frac{1}{s} \frac{e^{-as}}{1 + sq},$$

where a and q are chosen to give the required response (see Figure 9.7). If we let $a = kT$, then the z-transform of the output is

$$Y(z) = \frac{z^{-k-1}(1 - e^{-T/q})}{(1 - z^{-1})(1 - e^{-T/q}z^{-1})}$$

and the required transfer function is

$$T(z) = \frac{Y(z)}{R(z)} = \frac{z^{-k-1}(1 - e^{-T/q})}{(1 - z^{-1})(1 - e^{-T/q}z^{-1})} \frac{(1 - z^{-1})}{1}$$

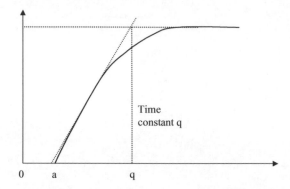

Figure 9.7 Dahlin controller response

or

$$T(z) = \frac{z^{-k-1}(1 - e^{-T/q})}{1 - e^{-T/q}z^{-1}}.$$

Using (9.3), we can find the transfer function of the required controller:

$$D(z) = \frac{1}{HG(z)} \frac{T(z)}{1 - T(z)} = \frac{1}{HG(Z)} \frac{z^{-k-1}(1 - e^{-T/q})}{1 - e^{-T/q}z^{-1} - (1 - e^{-T/q})z^{-k-1}}.$$

An example is given below to illustrate the use of the Dahlin controller.

Example 9.2

The open-loop transfer function of a plant is given by

$$G(s) = \frac{e^{-2s}}{1 + 10s}.$$

Design a Dahlin digital controller for the system. Assume that $T = 1$ s.

Solution

The transfer function of the system with a zero-order hold is given by

$$HG(z) = Z\left\{\frac{1 - e^{-sT}}{s}G(s)\right\} = (1 - z^{-1})Z\left\{\frac{e^{-2s}}{s(1 + 10s)}\right\}$$

or

$$HG(z) = (1 - z^{-1})z^{-2}Z\left\{\frac{1}{s(1 + 10s)}\right\} = (1 - z^{-1})z^{-2}Z\left\{\frac{1/10}{s(s + 1/10)}\right\}.$$

From z-transform tables we obtain

$$HG(z) = (1 - z^{-1})z^{-2}\frac{z(1 - e^{-0.1})}{(z - 1)(z - e^{-0.1})} = z^{-3}\frac{(1 - e^{-0.1})}{1 - e^{-0.1}z^{-1}}$$

or

$$HG(z) = \frac{0.095z^{-3}}{1 - 0.904z^{-1}}.$$

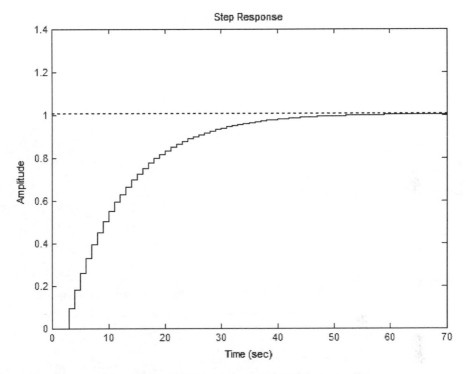

Figure 9.8 System response with Dahlin controller

For the controller, if we choose $q = 10$, then

$$D(z) = \frac{1}{HG(z)} \frac{T(z)}{1 - T(z)} = \frac{1 - 0.904z^{-1}}{0.095z^{-3}} \frac{z^{-k-1}(1 - e^{-0.1})}{1 - e^{-0.1}z^{-1} - (1 - e^{-0.1})z^{-k-1}}$$

or

$$D(z) = \frac{1 - 0.904z^{-1}}{0.095z^{-3}} \frac{0.095z^{-k-1}}{1 - 0.904z^{-1} - 0.095z^{-k-1}}$$

For realizability, if we choose $k = 2$, we obtain

$$D(z) = \frac{0.095z^3 - 0.0858z^2}{0.095z^3 - 0.0858z^2 - 0.0090}.$$

Figure 9.8 shows the step response of the system. It is clear that the response is exponential as expected.

The response of the controller is shown in Figure 9.9. Although the system response is slower, the controller signal is more acceptable.

9.1.3 Pole-Placement Control – Analytical

The response of a system is determined by the positions of its closed-loop poles. Thus, by placing the poles at the required points we should be able to control the response of a system.

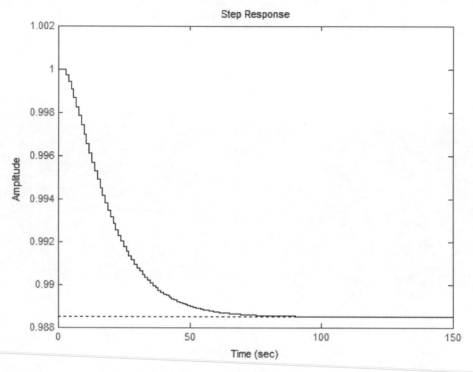

Figure 9.9   Controller response

Given the pole positions of a system, (9.3) gives the required transfer function of the controller as

$$D(z) = \frac{1}{HG(z)} \frac{T(z)}{1 - T(z)}.$$

$T(z)$ is the required transfer function, which is normally in the form of a polynomial. The denominator of $T(z)$ is constructed from the positions of the required roots. The numerator polynomial can then be selected to satisfy certain criteria in the system. An example is given below.

Example 9.3

The open-loop transfer function of a system together with a zero-order hold is given by

$$HG(z) = \frac{0.03(z + 0.75)}{z^2 - 1.5z + 0.5}.$$

Design a digital controller so that the closed-loop system will have $\zeta = 0.6$ and $w_d = 3$ rad/s. The steady-state error to a step input should be zero. Also, the steady-state error to a ramp input should be 0.2. Assume that $T = 0.2$ s.

Solution

The roots of a second-order system are given by

$$z_{1,2} = e^{-\zeta \omega_n T \pm j \omega_n T \sqrt{1 - \zeta^2}} = e^{-\zeta w_n T} (\cos \omega_n T \sqrt{1 - \zeta^2} \pm j \sin \omega_n T \sqrt{1 - \zeta^2}).$$

Thus, the required pole positions are

$$z_{1,2} = e^{-0.6 \times 3.75 \times 0.2}(\cos(0.2 \times 3) \pm j \sin(0.2 \times 3)) = 0.526 \pm j0.360.$$

The required controller then has the transfer function

$$T(z) = \frac{b_0 + b_1 z^{-1} + b_2 z^{-2} + b_3 z^{-3} + \cdots}{(z - 0.526 + j0.360)(z - 0.526 - j0.360)}$$

which gives

$$T(z) = \frac{b_0 + b_1 z^{-1} + b_2 z^{-2} + b_3 z^{-3} + \cdots}{1 - 1.052 z^{-1} + 0.405 z^{-2}}. \tag{9.6}$$

We now have to determine the parameters of the numerator polynomial. To ensure realizability, $b_0 = 0$ and the numerator must only have the b_1 and b_2 terms. Equation (9.6) then becomes

$$T(z) = \frac{b_1 z^{-1} + b_2 z^{-2}}{1 - 1.052 z^{-1} + 0.405 z^{-2}}. \tag{9.7}$$

The other parameters can be determined from the steady-state requirements.

The steady-state error is given by

$$E(z) = R(z)[1 - T(z)].$$

For a unit step input, the steady-state error can be determined from the final value theorem, i.e.

$$E_{ss} = \lim_{z \to 1} \frac{z - 1}{z} \frac{z}{z - 1}[v]$$

or

$$E_{ss} = 1 - T(1). \tag{9.8}$$

From (9.8), for a zero steady-state error to a step input,

$$T(1) = 1$$

From (9.7), we have

$$T(1) = \frac{b_1 + b_2}{0.353} = 1$$

or

$$b_1 + b_2 = 0.353, \tag{9.9}$$

and

$$T(z) = \frac{b_1 z + b_2}{z^2 - 1.052 z + 0.405}. \tag{9.10}$$

If K_v is the system velocity constant, for a steady-state error to a ramp input we can write

$$E_{ss} = \lim_{z \to 1} \frac{(z - 1)}{z} \frac{T z}{(z - 1)^2}[1 - T(z)] = \frac{1}{K_v}$$

or, using L'Hospital's rule,

$$\left. \frac{dT}{dz} \right|_{z=1} = -\frac{1}{K_v T}.$$

Thus from (9.10),

$$\left.\frac{dT}{dz}\right|_{z=1} = \frac{b_1(z^2 - 1.052z + 0.405) - (b_1z + b_2)(2z - 1.052)}{(z^2 - 1.052z + 0.405)^2} = -\frac{1}{K_vT} = -\frac{0.2}{0.2} = -1,$$

giving

$$\frac{0.353b_1 - (b_1 + b_2)0.948}{0.353^2} = -1$$

or

$$0.595b_1 + 0.948b_2 = 0.124, \tag{9.11}$$

From (9.9) and (9.11) we obtain,

$$b_1 = 0.596 \text{ and } b_2 = -0.243.$$

Equation (9.10) then becomes

$$T(z) = \frac{0.596z - 0.243}{z^2 - 1.052z + 0.405}. \tag{9.12}$$

Equation (9.12) is the required transfer function. We can substitute in Equation (9.3) to find the transfer function of the controller:

$$D(z) = \frac{1}{HG(z)} \frac{T(z)}{1 - T(z)} = \frac{z^2 - 1.5z + 0.5}{0.03(z + 0.75)} \frac{T(z)}{1 - T(z)}$$

or,

$$D(z) = \frac{z^2 - 1.5z + 0.5}{0.03(z + 0.75)} \frac{0.596z - 0.243}{z^2 - 1.648z + 0.648}$$

which can be written as

$$D(z) = \frac{0.596z^3 - 1.137z^2 + 0.662z - 0.121}{0.03z^3 - 0.027z^2 - 0.018z + 0.015} \tag{9.13}$$

The step response of the system with the controller is shown in Figure 9.10.

9.1.4   Pole-Placement Control – Graphical

In the previous subsection we saw how the response of a closed-loop system can be shaped by placing its poles at required points in the z-plane. In this subsection we will be looking at some examples of pole placement using the root-locus graphical approach.

When it is required to place the poles of a system at required points in the z-plane we can either modify the gain of the system or use a dynamic compensator (such as a phase lead or a phase lag). Given a first-order system, we can modify only the d.c. gain to achieve the required time constant. For a second-order system we can generally modify the d.c. gain to achieve a constant damping ratio greater than or less than a required value, and, depending on the system, we may also be able to design for a required natural frequency by simply varying the d.c. gain. For more complex requirements, such as placing the system poles at specific points in the z-plane, we will need to use dynamic compensators, and a simple gain adjustment alone

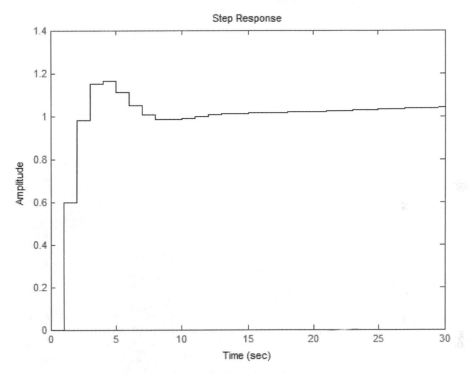

Figure 9.10   Step response of the system

will not be adequate. Some example pole-placement techniques are given below using the root locus approach.

Example 9.4

The block diagram of a sampled data control system is shown in Figure 9.11. Find the value of d.c. gain K which yields a damping ratio of $\zeta = 0.7$.

Solution

In this example, we will draw the root locus of the system as the gain K is varied, and then we will superimpose the lines of constant damping ratio on the locus. The value of K for the required damping ratio can then be read from the locus.

The root locus of the system is shown in Figure 9.12. The locus has been expanded for clarity between the real axis points $(-1, 1)$ and the imaginary axis points $(-1, 1)$, and the lines of constant damping ratio are shown in Figure 9.13. A vertical and a horizontal line are

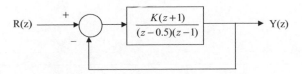

Figure 9.11   Block diagram for Example 9.4

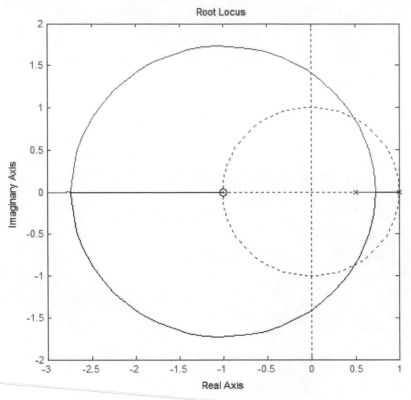

Figure 9.12 Root locus of the system

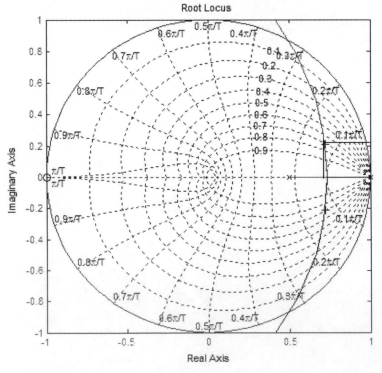

Figure 9.13 Root locus with the lines of constant damping ratio

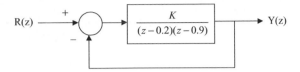

Figure 9.14   Block diagram for Example 9.5

drawn from the point where the damping factor is 0.7. At the required point the roots are $z_{1,2} = 0.7191 \pm j0.2114$. The value of K at this point is calculated to be $K = 0.0618$.

In this example, the required specification was obtained by simply modifying the d.c. gain of the system. A more complex example is given below where it is required to place the poles at specific points in the z-plane.

Example 9.5

The block diagram of a digital control system is given in Figure 9.14. It is required to design a controller for this system such that the system poles are at the points $z_{1,2} = 0.3 \pm j0.3$.

Solution

In this example, we will draw the root locus of the system and then use a dynamic compensator to modify the shape of the locus so that it passes through the required points in the z-plane.

The root locus of the system without the compensator is shown in Figure 9.15. The point where we want the roots to be is marked with a $\times$ and clearly the locus will not pass through this point by simply modifying the d.c. gain K of the system.

The angle of $G(z)$ at the required point is

$$\angle G(z) = -\angle(0.3 + j0.3 - 0.2) - \angle(0.3 + j0.3 - 0.9)$$

or

$$\angle G(z) = -\tan^{-1}\frac{0.3}{0.1} - \tan^{-1}\frac{0.3}{-0.6} = -45°.$$

Since the sum of the angles at a point in the root locus must be a multiple of $-180°$, the compensator must introduce an angle of $-180 - (-45) = -135°$. The required angle can be obtained using a compensator with a transfer function

$$D(Z) = \frac{z - n}{z - p}.$$

The angle introduced by the compensator is

$$\angle D(Z) = \angle(0.3 + j0.3 - n) - \angle(0.3 + j0.3 - p) = -135°$$

or

$$\tan^{-1}\frac{0.3}{0.3 - n} - \tan^{-1}\frac{0.3}{0.3 - p} = -135°.$$

There are many combinations of p and n which will give the required angle. For example, if we choose $n = 0.5$, then,

$$124° - \tan^{-1}\frac{0.3}{0.3 - p} = -135°$$

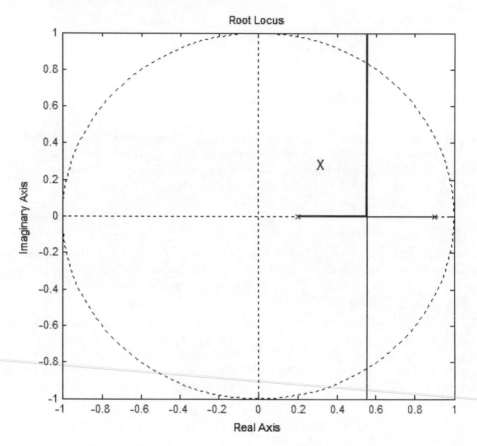

Figure 9.15   Root locus of the system without compensator

or

$$p = 0.242.$$

The required controller transfer function is then

$$D(z) = \frac{z - 0.5}{z - 0.242}.$$

The compensator introduces a zero at $z = 0.5$ and a pole at $z = 0.242$. The root locus of the compensated system is shown in Figure 9.16. Clearly the new locus passes through the required points $z_{1,2} = 0.3 \pm j0.3$, and it will be at these points that the d.c. gain is $K = 0.185$. The step response of the system with the compensator is shown in Figure 9.17. It is clear from this diagram that the system has a steady-state error.

The block diagram of the controller and the system is given in Figure 9.18.

Example 9.6

The block diagram of a system is as shown in Figure 9.19. It is required to design a controller for this system with percent overshoot (PO) less than 17 % and settling time $t_s \leq 10$ s. Assume that $T = 0.1$ s.

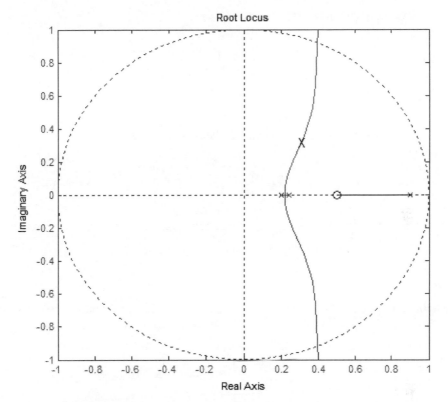

Figure 9.16   Root locus of the compensated system

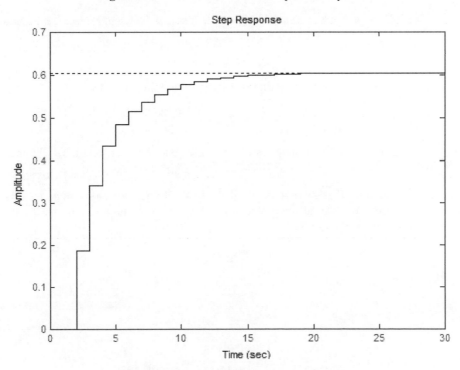

Figure 9.17   Step response of the system

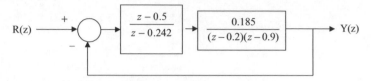

Figure 9.18   Block diagram of the controller and the system

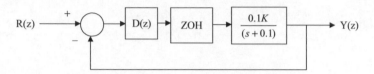

Figure 9.19   Block diagram for Example 9.6

Solution

The damping ratio, natural frequency and hence the required root positions can be determined as follows:

$$\text{For } PO < 17\%, \quad \zeta \geq 0.5.$$

$$\text{For } t_s \leq 10, \quad \zeta\omega_n \geq \frac{4.6}{t_s} \quad \text{or} \quad \omega_n \geq 0.92 \,\text{rad/s}.$$

Hence, the required pole positions are found to be

$$z_{1,2} = e^{-\zeta\omega_n T} \left(\cos \omega_n T \sqrt{1-\zeta^2} + j \sin \omega_n T \sqrt{1-\zeta^2} \right)$$

or

$$z_{1,2} = 0.441 \pm j0.451.$$

The z-transform of the plant, together with the zero-order hold, is given by

$$G(z) = \frac{z-1}{z} Z\left[\frac{0.1K}{s^2(s+0.1)} \right] = \frac{0.00484K(z+0.9672)}{(z-1)(z-0.9048)}.$$

The root locus of the uncompensated system and the required root position is shown in Figure 9.20.

It is clear from the figure that the root locus will not pass through the marked point by simply changing the d.c. gain. We can design a compensator as in Example 9.5 such that the locus passes through the required point, i.e.

$$D(Z) = \frac{z-n}{z-p}.$$

The angle of $G(z)$ at the required point is

$$\angle G(z) = \angle 0.441 + j0.451 + 0.9672 - \angle(0.441 + j0.451 - 1) - \angle(0.441 + j0.451 - 0.9048)$$

or

$$\angle G(z) = \tan^{-1}\frac{0.451}{1.4082} - \tan^{-1}\frac{0.451}{-0.559} - \tan^{-1}\frac{0.451}{-0.4638} = -259°.$$

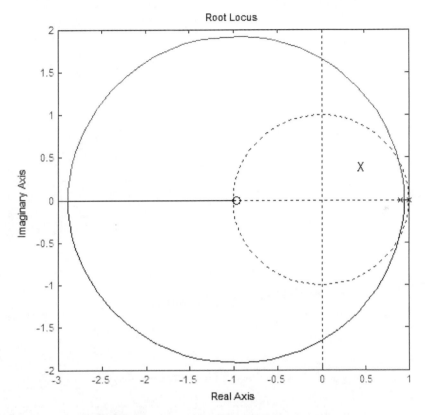

Figure 9.20 Root locus of uncompensated system

Since the sum of the angles at a point in root locus must be a multiple of $-180°$, the compensator must introduce an angle of $-180 - (-259) = 79°$. The required angle can be obtained using a compensator with a transfer function, and the angle introduced by the compensator is

$$\angle D(Z) = \angle(0.441 + j0.451 - n) - \angle(0.441 + j0.451 - p) = 79°$$

or

$$\tan^{-1}\frac{0.451}{0.441 - n} - \tan^{-1}\frac{0.451}{0.441 - p} = 79°.$$

If we choose $n = 0.6$, then

$$109° - \tan^{-1}\frac{0.451}{0.441 - p} = 79°$$

or

$$p = -0.340.$$

The transfer function of the compensator is thus

$$D(z) = \frac{z - 0.6}{z + 0.340}.$$

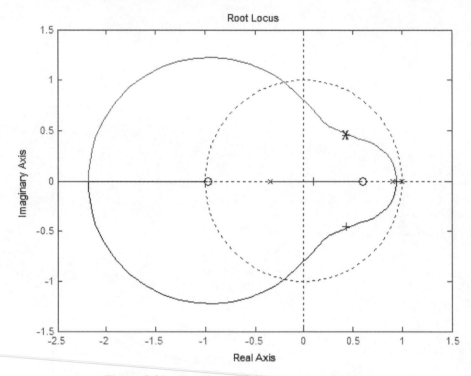

Figure 9.21   Root locus of the compensated system

Figure 9.21 shows the root locus of the compensated system. Clearly the locus passes through the required point. The d.c. gain at this point is $K = 123.9$.

The time response of the compensated system is shown in Figure 9.22.

9.2 PID CONTROLLER

The proportional–integral–derivative (PID) controller is often referred to as a 'three-term' controller. It is currently one of the most frequently used controllers in the process industry. In a PID controller the control variable is generated from a term proportional to the error, a term which is the integral of the error, and a term which is the derivative of the error.

- *Proportional*: the error is multiplied by a gain K_p. A very high gain may cause instability, and a very low gain may cause the system to drift away.

- *Integral*: the integral of the error is taken and multiplied by a gain K_i. The gain can be adjusted to drive the error to zero in the required time. A too high gain may cause oscillations and a too low gain may result in a sluggish response.

- *Derivative*: The derivative of the error is multiplied by a gain K_d. Again, if the gain is too high the system may oscillate and if the gain is too low the response may be sluggish.

Figure 9.23 shows the block diagram of the classical continuous-time PID controller. Tuning the controller involves adjusting the parameters K_p, K_d and K_i in order to obtain a satisfactory

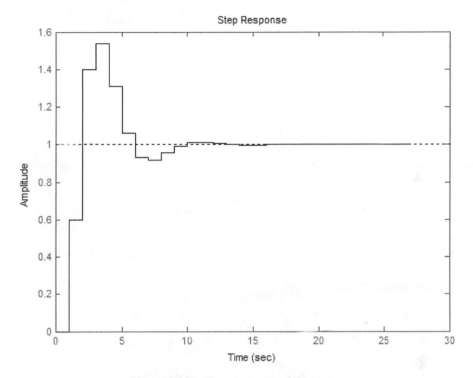

Figure 9.22   Time response of the system

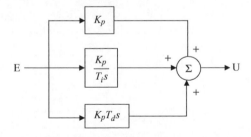

Figure 9.23   Continuous-time system PID controller

response. The characteristics of PID controllers are well known and well established, and most modern controllers are based on some form of PID.

The input–output relationship of a PID controller can be expressed as

$$u(t) = K_p \left[e(t) + \frac{1}{T_i} \int_0^t e(t)dt + T_d \frac{de(t)}{dt} \right], \tag{9.14}$$

where $u(t)$ is the output from the controller and $e(t) = r(t) - y(t)$, in which $r(t)$ is the desired set-point (reference input) and $y(t)$ is the plant output. T_i and T_d are known as the integral and derivative action time, respectively. Notice that (9.14) is sometimes written as

$$u(t) = K_p e(t) + K_i \int_0^t e(t)dt + K_d \frac{de(t)}{dt} + u_0, \tag{9.15}$$

where

$$K_i = \frac{K_p}{T_i} \quad \text{and} \quad K_d = K_p T_d. \tag{9.16}$$

Taking the Laplace transform of (9.14), we can write the transfer function of a continuous-time PID as

$$\frac{U(s)}{E(s)} = K_p + \frac{K_p}{T_i s} + K_p T_d s. \tag{9.17}$$

To implement the PID controller using a digital computer we have to convert (9.14) from a continuous to a discrete representation. There are several methods for doing this and the simplest is to use the trapezoidal approximation for the integral and the backward difference approximation for the derivative:

$$\frac{de(t)}{dt} \approx \frac{e(kT) - e(kT - T)}{T} \quad \text{and} \quad \int_0^t e(t)dt \approx \sum_{k=1}^n T e(kT).$$

Equation (9.14) thus becomes

$$u(kT) = K_p \left[e(kT) + T_d \frac{e(kT) - e(kT - T)}{T} + \frac{T}{T_i} \sum_{k=1}^n e(kT) \right] + u_0. \tag{9.18}$$

The PID given by (9.18) is now in a suitable form which can be implemented on a digital computer. This form of the PID controller is also known as the *positional* PID controller. Notice that a new control action is implemented at every sample time.

The discrete form of the PID controller can also be derived by finding the z-transform of (9.17):

$$\frac{U(z)}{E(z)} = K_p \left[1 + \frac{T}{T_i(1 - z^{-1})} + T_d \frac{(1 - z^{-1})}{T} \right]. \tag{9.19}$$

Expanding (9.19) gives

$$u(kT) = u(kT - T) + K_p[e(kT) - e(kT - T)] + \frac{K_p T}{T_i} e(kT)$$

$$+ \frac{K_p T_d}{T}[e(kT) - 2e(kT - T) - e(kT - 2T)]. \tag{9.20}$$

This form of the PID controller is known as the *velocity* PID controller. Here the current control action uses the previous control value as a reference. Because only a change in the control action is used, this form of the PID controller provides a smoother bumpless control when the error is small. If a large error exists, the response of the velocity PID controller may be slow, especially if the integral action time T_i is large.

The two forms of the PID algorithm, (9.18) and (9.20), may look quite different, but they are in fact similar to each other. Consider the positional controller (9.18). Shifting back one sampling interval, we obtain

$$u(kT - T) = K_p \left[e(kT - T) + T_d \frac{e(kT - T) - e(kT - 2T)}{T} + \frac{T}{T_i} \sum_{k=1}^{n-1} e(kT) \right] + u_0.$$

Subtracting from (9.18), we obtain the velocity form of the controller, as given by (9.20).

9.2.1   Saturation and Integral Wind-Up

In practical applications the output value of a control action is limited by physical constraints. For example, the maximum voltage output from a device is limited. Similarly, the maximum flow rate that a pump can supply is limited by the physical capacity of the pump. As a result of this physical limitation, the error signal does not return to zero and the integral term keeps adding up continuously. This effect is called integral wind-up (or integral saturation), and as a result of it long periods of overshoot can occur in the plant response. A simple example of what happens is the following. Suppose we wish to control the position of a motor and a large set-point change occurs, resulting in a large error signal. The controller will then try to reduce the error between the set-point and the output. The integral term will grow by summing the error signals at each sample and a large control action will be applied to the motor. But because of the physical limitation of the motor electronics the motor will not be able to respond linearly to the applied control signal. If the set-point now changes in the other direction, then the integral term is still large and will not respond immediately to the set-point request. Consequently, the system will have a poor response when it comes out of this condition.

The integral wind-up problem affects positional PID controllers. With velocity PID controllers, the error signals are not summed up and as a result integral wind-up will not occur, even though the control signal is physically constrained.

Many techniques have been developed to eliminate integral wind-up from the PID controllers, and some of the popular ones are as follows:

- Stop the integral summation when saturation occurs. This is also called conditional integration. The idea is to set the integrator input to zero if the controller output is saturated and the input and output are of the same sign.

- Fix the limits of the integral term between a minimum and a maximum.

- Reduce the integrator input by some constant if the controller output is saturated. Usually the integral value is decreased by an amount proportional to the difference between the unsaturated and saturated (i.e maximum) controller output.

- Use the velocity form of the PID controller.

9.2.2   Derivative Kick

Another possible problem when using PID controllers is caused by the derivative action of the controller. This may happen when the set-point changes sharply, causing the error signal to change suddenly. Under such a condition, the derivative term can give the output a *kick*, known as a *derivative kick*. This is usually avoided in practice by moving the derivative term to the feedback loop. The proportional term may also cause a sudden kick in the output and it is common to move the proportional term to the feedback loop.

9.2.3 PID Tuning

When a PID controller is used in a system it is important to tune the controller to give the required response. Tuning a PID controller involves selecting values for the controller parameters K_p, T_i and T_d. There are many techniques for tuning a controller, ranging from the first techniques described by J.G. Ziegler and N.B. Nichols (known as the Ziegler–Nichols tuning algorithm) in 1942 and 1943, to recent auto-tuning controllers. In this section we shall look at the tuning of PID controllers using the Ziegler–Nichols tuning algorithm.

Ziegler and Nichols suggested values for the PID parameters of a plant based on open-loop or closed-loop tests of the plant. According to Ziegler and Nichols, the open-loop transfer function of a system can be approximated with a time delay and a single-order system, i.e.

$$G(s) = \frac{K e^{-sT_D}}{1 + sT_1},\tag{9.21}$$

where T_D is the system time delay (i.e. transportation delay), and T_1 is the time constant of the system.

9.2.3.1 Open-Loop Tuning

For open-loop tuning, we first find the plant parameters by applying a step input to the open-loop system. The plant parameters K, T_D and T_1 are then found from the result of the step test as shown in Figure 9.24.

Ziegler and Nichols then suggest using the PID controller settings given in Table 9.1 when the loop is closed. These parameters are based on the concept of minimizing the integral of the absolute error after applying a step change to the set-point.

An example is given below to illustrate the method used.

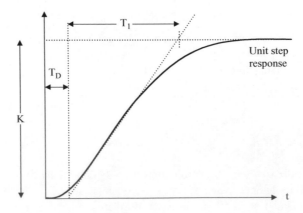

Figure 9.24   Finding plant parameters K, T_D and T_1

Table 9.1 Open-loop Ziegler–Nichols settings

Controller	K_p	T_i	T_d
Proportional	$\frac{T_1}{KT_D}$		
PI	$\frac{0.9T_1}{KT_D}$	$3.3T_D$	
PID	$\frac{1.2T_1}{KT_D}$	$2T_D$	$0.5T_D$

Example 9.7

The open-loop unit step response of a thermal system is shown in Figure 9.25. Obtain the transfer function of this system and use the Ziegler–Nichols tuning algorithm to design (a) a proportional controller, (b) to design a proportional plus integral (PI) controller, and (c) to design a PID controller. Draw the block diagram of the system in each case.

Solution

From Figure 9.25, the system parameters are obtained as $K = 40°C$, $T_D = 5$ s and $T_1 = 20$ s, and the transfer function of the plant is

$$G(s) = \frac{40e^{-5s}}{1 + 20s}.$$

Proportional controller. According to Table 9.1, the Ziegler–Nichols settings for a proportional controller are:

$$K_p = \frac{T_1}{KT_D}.$$

Thus,

$$K_p = \frac{20}{40 \times 5} = 0.1,$$

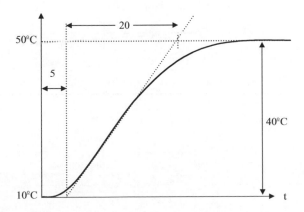

Figure 9.25 Unit step response of the system for Example 9.7

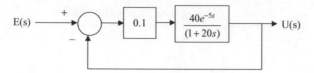

Figure 9.26   Block diagram of the system with proportional controller

The transfer function of the controller is then

$$\frac{U(s)}{E(s)} = 0.1,$$

and the block diagram of the closed-loop system with the controller is shown in Figure 9.26.

PI controller. According to Table 9.1, the Ziegler–Nichols settings for a PI controller are

$$K_p = \frac{0.9T_1}{KT_D} \quad \text{and} \quad T_i = 3.3T_D.$$

Thus,

$$K_p = \frac{0.9 \times 20}{40 \times 5} = 0.09 \quad \text{and} \quad T_i = 3.3 \times 5 = 16.5.$$

The transfer function of the controller is then

$$\frac{U(s)}{E(s)} = 0.09 \left[1 + \frac{1}{16.5s} \right] = \frac{0.09(16.5s + 1)}{16.5s}$$

and the block diagram of the closed-loop system with the controller is shown in Figure 9.27.

PID controller. According to Table 9.1, the Ziegler–Nichols settings for a PID controller are

$$K_p = \frac{1.2T_1}{KT_D}, \quad T_i = 2T_D, \quad T_d = 0.5T_D.$$

Thus,

$$K_p = \frac{1.2 \times 20}{40 \times 5} = 0.12, \quad T_i = 2 \times 5 = 10, \quad T_d = 0.5 \times 5 = 2.5.$$

The transfer function of the required PID controller is

$$\frac{U(s)}{E(s)} = K_p \left[1 + \frac{1}{T_i s} + Tds \right] = 0.12 \left[1 + \frac{1}{10s} + 2.5s \right]$$

or

$$\frac{U(s)}{E(s)} = \frac{3s^2 + 1.2s + 0.12}{10s}.$$

The block diagram of the system, together with the controller, is shown in Figure 9.28.

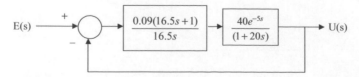

Figure 9.27   Block diagram of the system with PI controller

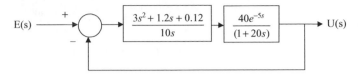

Figure 9.28   Block diagram of the system with PID controller

9.2.3.2   Closed-Loop Tuning

The Ziegler–Nichols closed-loop tuning algorithm is based on plant closed-loop tests. The procedure is as follows:

- Disable any derivative and integral action in the controller and leave only the proportional action.

- Carry out a set-point step test and observe the system response.

- Repeat the set-point test with increased (or decreased) controller gain until a stable oscillation is achieved (see Figure 9.29). This gain is called the *ultimate gain*, K_u.

- Read the period of the steady oscillation and let this be P_u.

- Calculate the controller parameters according to the following formulae: $K_p = 0.45K_u$, $T_i = P_u/1.2$ in the case of the PI controller; and $K_p = 0.6K_u$, $T_i = P_u/2$, $T_d = T_u/8$ in the case of the PID controller.

9.3   EXERCISES

1. The open-loop transfer function of a plant is given by:

$$G(s) = \frac{e^{-4s}}{1 + 2s}.$$

   (a) Design a dead-beat digital controller for the system. Assume that $T = 1$ s.
   (b) Draw the block diagram of the system together with the controller.
   (c) Plot the time response of the system.

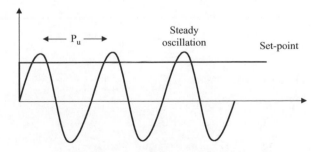

Figure 9.29   Ziegler–Nichols closed-loop test

2. Repeat Exercise 1 for a Dahlin controller. Plot the response and compare with the results obtained from the dead-beat controller.

3. The open-loop transfer function of a system together with a zero-order hold is given by

$$HG(z) = \frac{0.2(z + 0.8)}{z^2 - 1.5z + 0.5}.$$

Design a digital controller so that the closed-loop system will have $\zeta = 0.6$ and $w_d = 3$ rad/s. The steady-state error to a step input should be zero. Also, the steady state error to a ramp input should be 0.5. Assume that $T = 0.2$ s.

4. The block diagram of a sampled data control system is shown in Figure 9.30. Find the value of d.c. gain K to yield a damping ratio of 0.6

5. Draw the time response of the system in Exercise 4.

6. The open-loop transfer function of a system is

$$G(s) = \frac{K}{0.2s + 1}.$$

The system is preceded by a sampler and a zero-order hold. The closed-loop system is required to have a time constant of 0.4 s. (a) Determine the required value of the d.c. gain K. (b) Plot the unit step time response of the system with the controller.

7. The block diagram of a system is given in Figure 9.31. It is required to design a controller for this system such that the system poles are at the points $z_{1,2} = 0.4 \pm j0.4$ in the z-plane. (a) Derive the transfer function of the required digital controller. (b) Plot the unit step time response of the system without the controller. (c) Plot the unit step time response of the system with the controller.

8. The block diagram of a system is given in Figure 9.32. It is required to design a controller for this system with percent overshoot (PO) less than 20 % and settling time $t_s \leq 10$ s. Assume that the sampling time is, $T = 0.1$ s.
   (a) Derive the transfer function of the required digital controller.
   (b) Draw the block diagram of the system together with the controller.
   (c) Plot the unit step time response of the system without the controller.

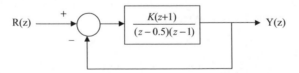

Figure 9.30   Block diagram for Exercise 4

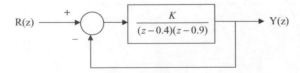

Figure 9.31   Block diagram for Exercise 7

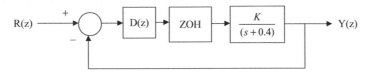

Figure 9.32   Block diagram for Exercise 8

(d) Plot the unit step time response of the system with the controller.

9. Explain the differences between the *position* and *velocity* forms of the PID controller.

10. The open-loop unit step response of a system is shown in Figure 9.33. Obtain the transfer function of this system and use the Ziegler–Nichols tuning algorithm to design:
(a) a proportional controller;
(b) a PI controller;
(c) a PID controller.
Draw the block diagram of the system in each case.

11. Explain the procedure for designing a PID controller using the Ziegler–Nichols algorithm when the plant is open-loop.

12. Repeat Exercise 11 for the case when the plant is closed-loop. What precautions should be taken when tests are performed on a closed-loop system?

13. Explain what integral wind-up is when a PID controller is used. How can integral wind-up be avoided?

14. Explain what derivative kick is when a PID controller is used. How can derivative kick be avoided?

15. The open-loop transfer function of a unity feedback system is

$$G(s) = \frac{10}{s(s + 10)}.$$

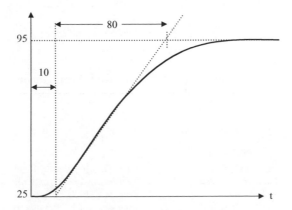

Figure 9.33   Unit step response of the system for Exercise 10

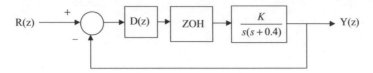

Figure 9.34   Block diagram for Exercise 17

Assume that $T = 1\,\mathrm{s}$ and design a controller so that the system response to a unit step input is

$$y(kT) = 0,\ 0.4,\ 1,\ 1,\ \ldots.$$

16. A mechanical process has the transfer function Ke^{-sT_D}/s The system oscillates with a frequency of $0.05\,\mathrm{Hz}$ when a unity gain feedback is applied. Determine the value of T_D.

17. The block diagram of a system is given in Figure 9.34. It is required to design a controller for this system with percent overshoot (PO) less than $15\,\%$ and settling time $t_s \le 10\,\mathrm{s}$. Assume that the sampling time is, $T = 0.2\,\mathrm{s}$.
    (a) Derive the transfer function of the required digital controller.
    (b) Draw the block diagram of the system together with the controller.
    (c) Plot the unit step time response of the system without the controller.
    (d) Plot the unit step time response of the system with the controller.

18. Derive an expression for the z-transform model of the continuous-time PID controller. Draw the block diagram of the controller. Describe how you can modify the model to avoid derivative kick.

19. The continuous-time PI controller has the transfer function

$$\frac{U(s)}{E(s)} = \frac{K_p s + K_i}{s}.$$

Derive the equivalent discrete-time controller transfer function using the bilinear transformation.

20. A commonly used compensator in the s-plane is the lead lag, or lag lead with transfer function

$$\frac{U(s)}{E(s)} = \frac{s + a}{s + b}.$$

Find the equivalent discrete-time controller using the bilinear transformation.

FURTHER READING

[Dorf, 1974]   Dorf, R.C. Modern Control Systems. Addison-Wesley Reading, MA, 1974.
[Franklin et al., 1990]   Franklin, G.F., Powell, J.D., and Workman, M.L. Digital Control of Dynamic Systems. 2nd edn., Addison-Wesley, Reading, MA, 1990.
[Katz, 1981]   Katz, P. Digital Control Using Microprocessors. Prentice Hall International, Englewood Cliffs, NJ, 1981.

[Kuo, 1963]   Kuo, B.C. Analysis and Synthesis of Sampled-Data Control Systems. Prentice Hall
                               International, Eaglewood Cliffs, NJ, 1963.

[Ogata, 1970]   Ogata, K. Modern Control Engineering. McGraw-Hill, New York, 1970.

[Phillips and Harbor, 1988]   Phillips, C.L. and Harbor, R.D. Feedback Control Systems. Prentice Hall, Eaglewood
                               Cliffs, NJ, 1988.

[Strum and Kirk, 1988]   Strum, R.D. and Kirk, D.E. Discrete Systems and Digital Signal Processing. Addison-
                               Wesley, Reading, MA, 1988.

[Tustin, 1947]   Tustin, A. A method of analyzing the behaviour of linear systems in terms of time
                               series. *J. Inst. Elect. Engineers.* **94**, Pt. IIA, 1947, pp. 130–142.

10

Controller Realization

A control algorithm which takes the form of a z-transform polynomial must be realized in the computer in the form of a program containing unit delays, constant multipliers, and adders.

A given controller transfer function can be realized in many different ways. Mathematically the alternative realizations are all equivalent, differing only in the way they are implemented. Different realizations have different computational efficiencies, different sensitivities to parameter errors, and different programming efforts are needed in each case. Only some of the important realizations, such as the direct structure, cascaded structure and parallel structure, as well as the second-order structures, are described in this chapter.

10.1 DIRECT STRUCTURE

The transfer function $D(z)$ of a digital controller can be represented in general by a ratio of two polynomials

$$D(z) = \frac{U(z)}{E(z)} = \frac{\sum_{j=0}^{n} a_j z^{-j}}{1 + \sum_{j=1}^{n} b_j z^{-j}}. \tag{10.1}$$

In direct structure the coefficients a_j and b_j appear as multipliers. There are several forms of direct structure, and we shall look at two of the most popular ones: the direct canonical structure and the direct noncanonical structure.

10.1.1 Direct Canonical Structure

Remembering that $b_0 = 1$, we can rewrite (10.1) as

$$D(z) = \frac{U(z)}{E(z)} = \frac{\sum_{j=0}^{n} a_j z^{-j}}{1 + \sum_{j=0}^{n} b_j z^{-j}}. \tag{10.2}$$

Microcontroller Based Applied Digital Control   D. Ibrahim
© 2006 John Wiley & Sons, Ltd

Let us now introduce a variable $R(z)$ such that

$$\frac{U(z)}{R(z)}\frac{R(z)}{E(z)} = \frac{\sum\limits_{j=0}^{n} a_j z^{-j}}{\sum\limits_{j=0}^{n} b_j z^{-j}} \tag{10.3}$$

or

$$\frac{U(z)}{R(z)} = \sum\limits_{j=0}^{n} a_j z^{-j} \tag{10.4}$$

and

$$\frac{E(z)}{R(z)} = \sum\limits_{j=0}^{n} b_j z^{-j}. \tag{10.5}$$

Assume that the transfer function of a digital controller is

$$R(z) = E(z) - \sum\limits_{j=1}^{n} b_j z^{-j} R(z). \tag{10.6}$$

We can rewrite (10.4) as

$$U(z) = \sum\limits_{j=0}^{n} a_j z^{-j} R(z). \tag{10.7}$$

Equations (10.6) and (10.7) can be written in the time domain as

$$r_k = e_k - \sum\limits_{j=1}^{n} b_j r_{k-j} \tag{10.8}$$

and

$$u_k = \sum\limits_{j=0}^{n} a_j r_{k-j}. \tag{10.9}$$

Equations (10.8) and (10.9) define the direct form, and the block diagram of the implementation is shown in Figure 10.1. The controller is made up of delays, adders and multipliers. An example is given below.

Example 10.1

The transfer function of a digital controller is found to be

$$D(z) = \frac{1 + 2z^{-1} + 4z^{-2}}{1 + 2z^{-1} + 5z^{-2}}.$$

Draw the block diagram of the direct canonical realization of this controller.

Solution

With reference to (10.8), (10.9) and Figure 10.1, we can draw the required block diagram as in Figure 10.2.

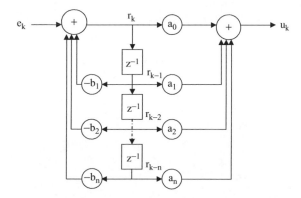

Figure 10.1   Canonical direct structure

10.1.2 Direct Noncanonical Structure

Consider Equation (10.2) with $b_0 = 1$:

$$D(z) = \frac{U(z)}{E(z)} = \frac{\displaystyle\sum_{j=0}^{n} a_j z^{-j}}{\displaystyle\sum_{j=0}^{n} b_j z^{-j}}$$

Cross multiplying and rewriting this equation we obtain

$$U(z) \sum_{j=0}^{n} b_j z^{-j} = E(z) \sum_{j=0}^{n} a_j z^{-j} \tag{10.10}$$

or, since $b_0 = 1$,

$$U(z) = \sum_{j=0}^{n} a_j z^{-j} E(z) - \sum_{j=1}^{n} b_j z^{-j} U(z). \tag{10.11}$$

Writing (10.11) in the time domain, we obtain the noncanonical form of the direct realization

$$u_k = \sum_{j=0}^{n} a_j e_{k-j} - \sum_{j=1}^{n} b_j u_{k-j}. \tag{10.12}$$

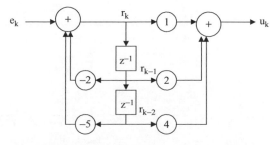

Figure 10.2   Block diagram for Example 10.1

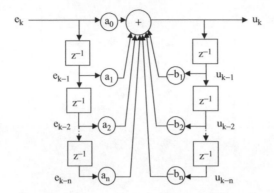

Figure 10.3 Noncanonical direct structure

The block diagram of the noncanonical direct realization is shown in Figure 10.3. This structure has only one adder, but $2n$ delay elements.

Example 10.2

The transfer function of a digital controller is found to be

$$D(z) = \frac{1 + 2z^{-1} + 4z^{-2}}{1 + 2z^{-1} + 5z^{-2}}.$$

Draw the block diagram of the direct noncanonical realization of this controller.

Solution

With reference to (10.12) and Figure 10.3, we can draw the required block diagram as in Figure 10.4.

10.2 CASCADE REALIZATION

The cascade realization is less sensitive to coefficient sensitivity problems. In this method the transfer function is implemented as a product of first-order and second-order transfer functions. Thus, the controller transfer function is written as

$$D(z) = P(z) \prod_{i=1}^{m} Q_i(z) \quad \text{for } n \text{ odd} \tag{10.13}$$

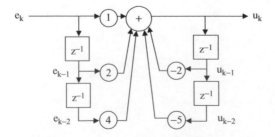

Figure 10.4 Block diagram for Example 10.2

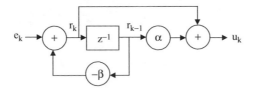

Figure 10.5 Realization of $P(z)$

and

$$D(z) = \prod_{i=1}^{m} Q_i(z) \quad \text{for } n \text{ even,}$$ (10.14)

where m is the smallest integer greater than or equal to $n/2$. $P(z)$ in (10.13) is the first-order transfer function

$$P(z) = \frac{1 + \alpha z^{-1}}{1 + \beta z^{-1}},$$ (10.15)

shown in Figure 10.5. With reference to Figure 10.5, we can write

$$r_k = e_k - \beta r_{k-1}$$ (10.16)

and

$$u_k = r_k + \alpha r_{k-1}.$$ (10.17)

$Q(z)$ in (10.13) and (10.14) is a second-order transfer function,

$$Q(z) = \frac{a_0 + a_1 z^{-1} + a_2 z^{-2}}{1 + b_1 z^{-1} + b_2 z^{-2}},$$ (10.18)

shown in Figure 10.6. With reference to Figure 10.6, we can write

$$r_k = e_k - b_1 r_{k-1} - b_2 r_{k-2}$$ (10.19)

and

$$u_k = a_0 r_k + a_1 r_{k-1} + a_2 r_{k-2}.$$ (10.20)

In practice, in order to avoid coefficient sensitivity problems second-order transfer function modules of the form given by (10.18) are frequently used and the modules are cascaded to give the required order. The block diagram of Figure 10.6 is sometimes drawn vertically, as shown in Figure 10.7.

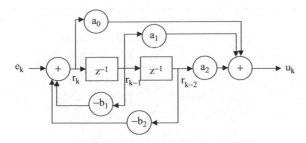

Figure 10.6 Realization of $Q(z)$

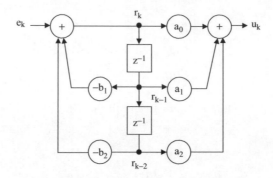

Figure 10.7   Figure 10.6 drawn vertically

Example 10.3

The transfer function of a digital controller is given by

$$D(z) = \frac{3(z+1)(z+2)}{z^2 + 0.4z + 0.03}.$$

Use two-first order cascaded transfer functions to implement this controller.

Solution

The transfer function can be factorized as

$$D(z) = \frac{3(z+1)(z+2)}{(z+0.1)(z+0.3)} = \frac{3(1+z^{-1})(1+2z^{-1})}{(1+0.1z^{-1})(1+0.3z^{-1})}.$$

The required cascaded realization is shown in Figure 10.8.

Example 10.4

The transfer function of a digital controller is given by

$$D(z) = \frac{(1 + 0.6z^{-1})(1 + 2z^{-1} + 4z^{-2})}{(1 + 0.4z^{-1})(1 + 0.1z^{-1} + 0.3z^{-2})}.$$

Use a first-order and a second-order cascaded transfer function to implement this controller.

Solution

The transfer function can be implemented as a cascade of a first-order and a second-order function, as shown in Figure 10.9.

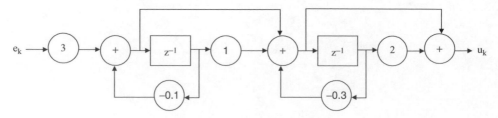

Figure 10.8   Cascaded realization for Example 10.3

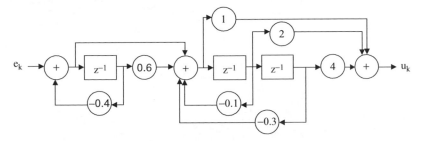

Figure 10.9   Cascaded realization for Example 10.4

10.3   *PARALLEL REALIZATION*

The parallel realization also avoids the coefficient sensitivity problem. In this method the transfer function is factored and written as a sum of first-order and second-order transfer functions:

$$D(z) = \alpha_0 + D_1(z) + D_2(z) + \cdots + D_m(z), \quad 1 < m < n.$$

First-order transfer functions are of the form

$$D_1(z) = \frac{\alpha}{1 + \beta z^{-1}}, \tag{10.21}$$

as shown in Figure 10.10. With reference to this figure, we can write

$$r_k = e_k - \beta r_{k-1} \tag{10.22}$$

and

$$u_k = \alpha r_k. \tag{10.23}$$

Second-order transfer functions are of the form

$$D_2(z) = \frac{a_1 + a_2 z^{-1}}{1 + b_1 z^{-1} + b_2 z^{-2}}, \tag{10.24}$$

as shown in Figure 10.11. With reference to this figure, we can write

$$r_k = e_k - b_1 r_{k-1} - b_2 r_{k-2} \tag{10.25}$$

and

$$u_k = a_0 r_k + a_1 r_{k-1} + r_{k-2}. \tag{10.26}$$

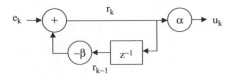

Figure 10.10   First-order element for parallel realization

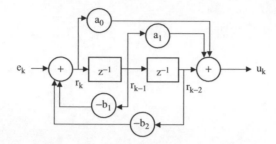

Figure 10.11   Second-order element for parallel realization

Example 10.5

The transfer function of a digital controller is given by

$$D(z) = \frac{(1 + z^{-1})(1 + 2z^{-1})}{(1 + 3z^{-1})(1 + 4z^{-1})}.$$

Realize this transfer function using first-order parallel transfer functions.

Solution

The controller transfer function can be factorized as follows:

$$D(z) = \frac{(1 + z^{-1})(1 + 2z^{-1})}{(1 + 3z^{-1})(1 + 4z^{-1})} = \frac{A}{1 + 3z^{-1}} + \frac{B}{1 + 4z^{-1}} + C.$$

From the partial fraction expansion, we obtain $A = -\frac{2}{3}$, $B = \frac{3}{2}$ and $C = \frac{1}{6}$. Thus,

$$D(z) = -\frac{2}{3(1 + 3z^{-1})} + \frac{3}{2(1 + 4z^{-1})} + \frac{1}{6}.$$

With reference to Figure 10.10, the controller implementation is shown in Figure 10.12.

10.4 PID CONTROLLER IMPLEMENTATIONS

PID controllers are very important in many process control applications. In this section we shall look at the realization of this type of controller.

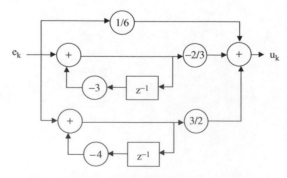

Figure 10.12   Realization for Example 10.4

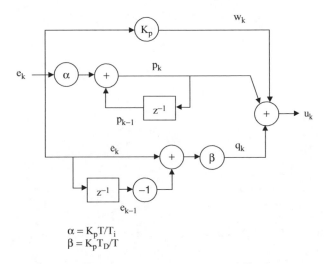

$$\alpha = K_pT/T_i$$
$$\beta = K_pT_D/T$$

Figure 10.13 PID controller as a parallel structure

The z-transform of the PID controller was derived in (9.19), and is reproduced here for convenience:

$$D(z) = K_p + \frac{K_pT}{T_i(1 - z^{-1})} + \frac{K_pT_d(1 - z^{-1})}{T}. \tag{10.27}$$

As shown in Figure 10.13, this transfer function may be implemented as a parallel structure by summing the proportional, integral and derivative terms.

With reference to Figure 10.13 we can write the following difference equations: for the proportional section,

$$w_k = K_pe_k. \tag{10.28}$$

for the integral section,

$$p_k = \alpha e_k + p_{k-1}; \tag{10.29}$$

and for the derivative section,

$$q_k = \beta(e_k - e_{k-1}). \tag{10.30}$$

The output is given by

$$u_k = w_k + p_k + q_k. \tag{10.31}$$

An alternative implementation of the PID would be to find a second order transfer function for (10.27) and then use the direct structure to implement it. Equation (10.27) can be written as

$$
\begin{aligned}
D(z) &= \frac{K_p(1 - z^{-1}) + K_pT/T_i + (K_pT_d/T)(1 - z^{-1})^2}{1 - z^{-1}} \\
&= \frac{K_p - K_pz^{-1} + K_pT/T_i + K_pT_d/T + (K_pT_d/T)z^{-2} - 2(K_pT_d/T)z^{-1}}{1 - z^{-1}} \\
&= \frac{K_p + K_pT/T_i + K_pT_d/T - (K_p + 2K_pT_d/T)z^{-1} + (K_pT_d/T)z^{-2}}{1 - z^{-1}}
\end{aligned}
$$

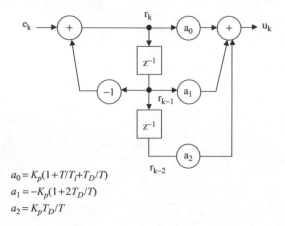

$$a_0 = K_p(1 + T/T_i + T_D/T)$$
$$a_1 = -K_p(1 + 2T_D/T)$$
$$a_2 = K_p T_D/T$$

Figure 10.14 PID implementation as a direct canonical structure

which is in the form

$$\frac{a_0 + a_1 z^{-1} + a_2 z^{-2}}{1 + b_1 z^{-1} + b_2 z^{-2}},$$ (10.32)

where

$$a_0 = K_p(1 + T/T_i + T_d/T), \quad a_1 = -K_p(1 + 2T_d/T), \quad a_2 = K_p T_d/T$$

and

$$b_1 = -1, \quad b_2 = 0.$$

Figure 10.14 shows the implementation of the PID controller as a second-order direct canonical structure.

The velocity form of the PID controller is used frequently in practice and the difference equation for this controller was derived in Chapter 9 – see (9.20). Considering this equation again and replacing kT simply by subscript k, we can write

$$u_k = u_{k-1} + K_p[e_k - e_{k-1}] + \frac{K_p T}{T_i} e_k + \frac{K_p T_d}{T}[e_k - 2e_{k-1} - e_{k-2}]$$ (10.33)

or

$$u_k = u_{k-1} + \left[K_p + \frac{K_p T}{T_i} + \frac{K_p T_d}{T} \right] e_k - \left[K_p + \frac{2K_p T_d}{T} \right] e_{k-1} + \frac{K_p T_d}{T} e_{k-2}.$$

Alternatively, we can write this in a simpler form as

$$u_k = u_{k-1} + a e_k + b e_{k-1} + c e_{k-2}$$ (10.34)

where

$$a = K_p + \frac{K_p T}{T_i} + \frac{K_p T_d}{T},$$

$$b = -\left[K_p + \frac{2K_p T_d}{T} \right],$$

$$c = \frac{K_p T_d}{T}.$$

By taking the z-transform of (10.34) we obtain

$$U(z) = z^{-1}U(z) + aE(z) + bz^{-1}E(z) + cz^{-2}E(z)$$

or

$$D(z) = \frac{U(z)}{E(z)} = \frac{a + bz^{-1} + cz^{-2}}{1 - z^{-1}}. \tag{10.35}$$

Equation (10.35) can easily be implemented using a direct realization. Notice that if only proportional plus integral (PI) action is required, the derivative constant T_d can be set to zero and we get the PI equation

$$D(z) = \frac{U(z)}{E(z)} = \frac{a + bz^{-1}}{1 - z^{-1}}, \tag{10.36}$$

with

$$a = K_p + \frac{K_p T}{T_i},$$

$$b = K_p.$$

Equation (10.36) can easily be implemented as a first-order transfer function.

10.5 MICROCONTROLLER IMPLEMENTATIONS

The final stage of a digital control system design is the implementation of the controller algorithm (set of difference equations) on a digital computer. In this section, we shall explore the implementation of digital controller algorithms on PIC microcontrollers. The PIC 16F877 microcontroller will be used in the examples since this microcontroller has a built-in A/D converter and a reasonable amount of program memory and data memory. There are many other microcontrollers in the PIC family with built-in A/D converters, and in general any of these can be used since the operation of microcontrollers in the PIC family with similar features is identical.

Microcontrollers have traditionally been programmed using the assembly language of the target hardware. Assembly language has several important disadvantages and is currently less popular than it used to be. One important disadvantage is that the code generated using the assembly language can only run on the specific target hardware. For example, the assembly program developed for a PIC microcontroller cannot be used, say, on an Intel 8051 micro-controller. Assembly language programs also tend to be more difficult to develop, test and maintain.

In this section, the Hi-Tech PICC language as described in Chapter 4 is used in the implementation of the algorithms. As described in Section 1.6 there are several methods that can be used to implement the controller algorithm. One of the most common, which has the advantage of accurate implementation, is the use of a timer interrupt to generate the required loop delay (or the sampling interval). In this method the software consists of two parts: the main program and the interrupt service routine. As shown in Figure 10.15(a) Figure 10.15, in the main program various variables, as well as the A/D converter and the timer interrupt mechanism, are initialized. The timer is set to interrupt at an interval equivalent to the sampling interval of the required digital controller. The main program then enters a loop waiting for the timer interrupts to occur. Whenever a timer interrupt occurs the program jumps the interrupt service

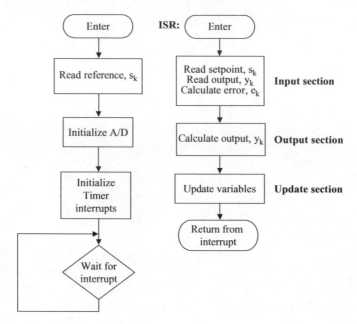

Figure 10.15 Controller implementation: (a) main program; (b) interrupt service routine

routine (ISR) as shown in Figure 10.15(b), and it is within this routine that the actual controller algorithm is implemented. The error signal is obtained by calculating the difference between the reference values and measured values. The algorithm is then implemented and the output sample for the current sampling time is obtained. A preprocessing step is then performed by updating the variables for the next sample. On return from the ISR, the program waits in the main program until the next sampling interval, and the above process repeats.

10.5.1 Implementing Second-Order Modules

In Section 10.2 we saw how a second-order module can be realized using adders, multipliers and delay elements. The second-order module is shown in Figure 10.16. The difference equations describing such a module are (10.19) and (10.20). If we let

$$M_1 = -b_1 r_{k-1} - b_2 r_{k-2}$$

and

$$M_2 = a_1 r_{k-1} + a_2 r_{k-2}$$

then the difference equations for the second-order module become

$$r_k = e_k + M_1,$$

$$u_k = a_0 r_k + M_2.$$

The implementation of the second-order module is shown as a flow-chart in Figure 10.17. This figure does not show the initialization of the variables T_1, T_2, r_1, r_2, the A/D initialization,

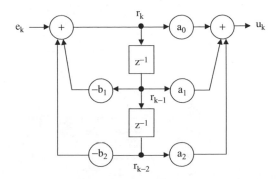

Figure 10.16   Second-order module implementation

the I/O port initialization, etc.; it only shows the interrupt service routine. A PIC microcontroller program for the implementation of a second-order module is given in the following example.

Example 10.6

The circuit diagram of a digital control system is shown in Figure 10.18. A PIC16F877 microcontroller is to be used as the digital controller in this system. Assume that the set-point input s is to be hard-coded to the program, and the output y is analog and connected to A/D channel AN0 (bit 0 of port A) of the microcontroller. The microcontroller is assumed to operate with a crystal frequency of 4 MHz as shown in the figure. The output (port B) of the microcontroller is interfaced to a D/A converter which acts as a zero-order-hold and generates an analog output to drive the plant.

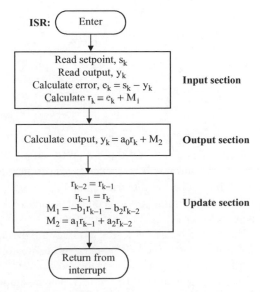

Figure 10.17   Flow diagram of second order module

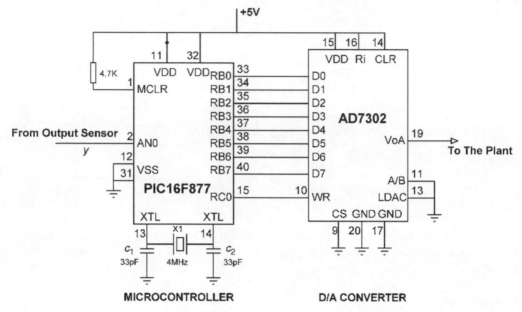

Figure 10.18   Circuit diagram of the microcontroller

Assume that the digital controller to be implemented is in the form of a second-order module, and write a program in C to implement this controller. The controller parameters are assumed to be

$$a_0 = 1, \quad a_1 = 0.8, \quad a_2 = 1.2, \quad b_1 = 1.85, \quad b_2 = 0.92,$$

i.e. the required controller transfer function is

$$D(z) = \frac{1 + 0.8z^{-1} + 1.2z^{-2}}{1 + 1.85z^{-1} + 0.92z^{-2}}.$$

Also assume that the required sampling interval is $T = 0.01$s.

Solution

The controller hardware is based on a PIC16F877 microcontroller. The microcontroller is operated from a 4 MHz crystal, connected to OSC1 and OSC2 inputs. With a 4 MHz crystal, the basic clock rate of the microcontroller is 1μs (the crystal frequency is divided by 4 to obtain the basic timing rate). The analog output of the plant (y) is connected to A/D converter channel AN0 of the microcontroller. Similarly, port B digital outputs of the microcontroller are connected to an AD7302 type D/A converter. The operation of this D/A converter is very simple. Digital data is applied to eight inputs D0–D7, while the write control input, WR, is at logic high. The analog data appears at the output after the WR input is lowered to logic low. The WR input of the D/A converter is controlled from port pin RC0 of the microcontroller. The D/A converter is set to operate with a full-scale reference voltage of +5 V. The resolution of the converter is 8 bits, i.e. there are $2^8 = 256$ quantization levels. With a full-scale reference voltage of +5 V, the resolution is $5000/256 = 19.53\,\text{mV}$, i.e. the digital bit pattern '00000001' corresponds to $19.53\,\text{mV}$, the bit pattern '00000010' corresponds to $2 \times 19.53 = 39.06\,\text{mV}$ and so on.

The first program attempt is shown in Figure 10.19. Comments are used to describe operation of various parts of the program. The program consists of the *main* program and the functions: *Initialize_AD*, *Initialize_Timer*, *Read_AD_Input*, and *ISR*.

Main program. The coefficients of the controller are defined at the beginning of the main program. Also, the A/D converter and the timer initialization functions are called here. The main program then enables global interrupts and enters an endless loop waiting for timer interrupts to occur.

ISR. This is the interrupt service routine. The program jumps to this function every 10 ms. The function reads a sample, and calculates the error term e_k. The output value y_k is then calculated and sent to the D/A converter. In the final part of the ISR, the variables are updated for the next sample, and the timer interrupt is re-enabled.

Initialize_Timer. This function initializes the timer TMR0 so that timer interrupts can be generated at 10 ms intervals. As described in Chapter 3, the timing interval depends on the clock frequency, the pre-scaler value, and the data loaded into the TMR0 register. It can be shown that the timing interval is given by:

$$\text{Timing\_interval} = 4*\text{clock\_period}*\text{prescaler}*(256 - \text{TMR0\_value})$$

and the value to be loaded into TMR0 register for a required timing interval is given by:

$$\text{TMR0\_value} = 256 - \text{timing\_interval}/(4*\text{clock\_period}*\text{prescaler})$$

The clock frequency is chosen as 4 MHz, i.e. the clock_period $= 0.25\,\mu\text{s}$. If we choose a pre-scaler value of 64, the value to be loaded into the timer register for a 10 ms (10 000 μs) timing interval can be calculated as:

$$\text{TMR0\_value} = 256 - 10000\,\text{ms}/(4*0.25*64) = 99.75$$

We can choose 100 as the nearest value. This will give a timing interval of

$$\text{Timing\_interval} = 4*0.25*64*(256 - 100) = 9.984\text{ms},$$

which is very close to the required value.

Thus, the *Initialize_Timer* function is programmed as follows:

```
/* This function initilizes the timer TMR0 so that interrupts can be
generated at every 10ms intervals */
void Initialize_Timer(void)
{
  T0CS = 0;           /* Select f/4 clock for the TMR0 */
  PSA = 0;            /* Select pre-scaler */
  PS0 = 1;            /* Set pre-scaler to 64 */
  PS1 = 0;            /* PS2,PS1,PS0 = 101 */
  PS2 = 1;
  TMR0 = 100;         /* Load TMR0 = 100 */
  T0IE = 1;           /* Enable TMR0 interrupts */
  T0IF = 0;           /* Clear TMR0 interrupt flag */
}
```

```
/**********************************************************************

                        DIGITAL CONTROLLER
                        ==================

This program implements a second-order digital controller module on a
PIC16F877 (or equivalent) microcontroller.  The microcontroller operates with
a 4MHz crystal.  The analog input AN0 of the microcontroller is connected to
the output sensor of the plant (y).  The PORT B output of the microcontroller
is connected to a AD7302 type D/A converter.  The WR input of the controller
is controlled from port pin RC0 of the microcontroller.  The sampling
interval is 0.01s (10ms) and the timer interrupt service routine is used to
obtain the required sampling interval.

Program : Second_Order_Module.C
Date : July 2005

**********************************************************************/
#include <pic.h>
#define DA_Write RC0

float DA_LSB,AD_LSB,a0,a1,a2, b1,b2,M1,M2,rk,rk_1,rk_2,ek,sk,yk,uk;

/* This function initializes the A/D converter so that analog data can be
received from channel AN0 of the microcontroller */
void Initialize_AD(void)
{
}

/* This function initilizes the timer TMR0 so that interrupts can be
generated at 10ms intervals */
void Initialize_Timer(void)
{
}

/* This function reads data from the A/D converter and stores in variable
yk */
void Read_AD_Input(void)
{
}

/* Interrupt Service Routine.  The program jumps here every 10 ms */
void interrupt ISR(void)
{
        Read_AD_Input();                    /* Read A/D input */

        ek = sk - yk;                       /* Calculate error term */
        rk = ek + M1;
        yk = a0*rk + M2;                    /* Calculate output */
        uk = yk*DA_LSB;
        PORTB = uk;                         /* Send to PORT B */
        DA_Write = 0;                        /* Write to D/A converter */
        DA_Write = 1;
```

Figure 10.19 First program attempt (*Continued*)

```
        rk_2 = rk_1;                         /* Update variables */
        rk_1 = rk;
        M1 = -b1*rk_1 - b2*rk_2;
        M2 = a1*rk_1 + a2*rk_2;

        T0IF = 0;                            /* Re-enable timer interrupts */
}

/* Main Program.  The main program initializes the variables, A/D converter,
   D/A converter etc. and then waits in an endless loop for timer interrupts
   to Occur every 10 ms */

main(void)
{
        a0 = 1; a1 = 0.8;        a2 = 1.2;
        b1 = 1.85;               b2 = 0.92;
        M1 = 0;        M2 = 0;        rk = 0;      rk_1 = 0;     rk_2 = 0;
        sk = 1.0;
        DA_LSB = 5000.0/1024.0;

        TRISA = 1;                   /* RA0 (AN0) is input */
        TRISB = 0;                   /* PORT B is output */
        TRISC = 0;                   /* RC0 is output */

        DA_Write = 1;                /* Disable D/A converter */

        Initialize_AD();             /* Initialize A/D converter */
        Initialize_Timer();          /* Initialize timer interrupts */
        ei();                        /* Enable interrupts */
        for(;;);                     /* Wait for an interrupt */
}
```

Figure 10.19 (*Continued*)

Initialize_AD. This function initializes the A/D converter so that analog data can be received from channel AN0 of the microcontroller. As described in Chapter 3, the A/D converter is initialized by first selecting the reference voltage and the output data format using the ADCON1 register, and then selecting the A/D converter clock using the ADCON0 register. Thus, channel AN0 is configured by loading 0x8E into the ADCON1 register, and 0x41 into the ADCON0 register. Thus, the *Initialize_AD* function is programmed as follows:

```
/* This function initializes the A/D converter so that analog data
can be received from channel AN0 of the microcontroller */
void Initialize_AD(void)

{
  ADCON1 = 0 × 8E;        /* Configure AN0 for +5V reference */
  ADCON0 = 0 × 41;        /* Select A/D converter clock */
}
```

Read_AD_Input. This function starts the A/D converter and reads a sample at the end of the conversion. The conversion is started by setting the GO/DONE bit of ADCON0. The program should then wait until the conversion is complete, which is indicated by the GO/DONE bit becoming zero. The high 2 bits of the converted data are in ADRESH, and the low 8 bits are in register ADRESL. The 10-bit converted data is extracted and stored in variable y_k.

The *Read_AD_Input* function is programmed as follows:

```
/* This function reads data from the A/D converter and stores in
variable yk */
void Read_AD_Input(void)

{
ADCON0 = 0 x 45;                     /* Start A/D conversion */
while(ADCON0 & 4) != 0);             /* Wait until conversion completes */
y_high = ADRESH;                     /* High 2 bytes of converted data */
y_low = ADRESL;                      /* Low byte of converted data */
yk = 256.0*y_high + y_low;           /* Converted data in yk */
yk = yk*AD_LSB;                      /* Sensor output in mV */
}
```

AD_LSB converts the A/D value to into millivolts.

The complete program is shown in Figure 10.20.

10.5.2 Implementing First-Order Modules

In Section 10.2 we saw how a first-order module can be realized using adders, multipliers and delay elements. The first-order module is shown in Figure 10.21. The difference equations describing a first-order module were found to be

$$r_k = e_k - b_1 r_{k-1}$$

and

$$u_k = a_0 r_k + a_1 r_{k-1}.$$

If we let

$$M_1 = -b_1 r_{k-1}$$

and

$$M_2 = a_1 r_{k-1},$$

then the difference equations for the first-order module becomes

$$r_k = e_k + M_1$$
$$u_k = a_0 r_k + M_2$$

The implementation of the first-order module is similar to the second-order module and an example is given below.

```
/**************************************************************************

                        DIGITAL CONTROLLER
                        ==================

This program implements a second-order digital controller module on a
PIC16F877 (or equivalent) microcontroller.  The microcontroller operates with
a 4MHz crystal.  The analog input AN0 of the microcontroller is connected to
the output sensor of the plant (y).  The PORT B output of the microcontroller
is connected to an AD7302 type D/A converter.  The WR input of the controller
is controlled from port pin RC0 of the microcontroller.  The sampling
interval is 0.01s (10ms) and the timer interrupt service routine is used to
obtain the required sampling interval.

Program : Second_Order_Module.C
Date : July 2005

**************************************************************************/
#include <pic.h>
#define DA_Write RC0

float DA_LSB,AD_LSB,a0,a1,a2, b1,b2,M1,M2,rk,rk_1,rk_2,ek,sk,yk,uk;
float y_high,y_low;

/* This function initializes the A/D converter so that analog data can be
received from channel AN0 of the microcontroller */
void Initialize_AD(void)
{
      ADCON1 = 0x8E;          /* Configure AN0 for +5V reference */
      ADCON0 = 0x41;          /* Select A/D clock and channel */
}

/* This function initilizes the timer TMR0 so that interrupts can be
generated at 10ms intervals */
void Initialize_Timer(void)
{
      T0CS = 0;               /* Select f/4 clock for TMR0 */
      PSA = 0;                /* Select pre-scaler */
      PS0 = 1;                /* Set pre-scaler to 64 */
      PS1 = 0;                /* PS2,PS1,PS0 = 101 */
      PS2 = 1;
      TMR0 = 100;             /* Load TMR0=100 */
      T0IE = 1;               /* Enable TMR0 interrupts */
      T0IF = 0;               /* Clear TMR0 interrupt flag */
*/
}

/* This function reads data from the A/D converter and stores in variable
yk */
void Read_AD_Input(void)
{
      ADCON0 = 0x45;                    /* Start A/D conversion */
      While((ADCON0 & 4) != 0);         /* Wait until conversion completes */
```

Figure 10.20 Complete program of the second-order controller (*Continued*)

```
        y_high = ADRESH;                /* High 2 bits of converted data */
        y_low = ADRESL;                 /* Low byte of converted data */
        yk = 256.0*y_high + y_low;      /* Converted data in yk */
        yk = yk*AD_LSB;                 /* Sensor output in mV */
}

/* Interrupt Service Routine.  The program jumps here every 10 ms */
void interrupt ISR(void)
{
        TMR0 = 100;                     /* Reload TMR0 */
        Read_AD_Input();                /* Read A/D input */

        ek = sk - yk;                   /* Calculate error term */
        rk = ek + M1;
        yk = a0*rk + M2;                /* Calculate output */
        uk = yk*DA_LSB;
        PORTB = uk;                     /* Send to PORT B */

        DA_Write = 0;                   /* Write to D/A converter */
        DA_Write = 1;

        rk_2 = rk_1;                    /* Update variables */
        rk_1 = rk;
        M1 = -b1*rk_1 - b2*rk_2;
        M2 = a1*rk_1 + a2*rk_2;

        T0IF = 0;                       /* Re-enable timer interrupts */
}

/* Main Program.  The main program initializes the variables, A/D converter,
    D/A converter etc. and then waits in an endless loop for timer
    interrupts to Occur every 10 ms */

main(void)
{
        a0 = 1;       a1 = 0.8;      a2 = 1.2;
        b1 = 1.85;    b2 = 0.92;
        M1 = 0;        M2 = 0;          rk = 0;    rk_1 = 0;   rk_2 = 0;
        sk = 1.0;
        AD_LSB = 5000.0/1024.0;
        DA_LSB = 256.0/5000.0;

        TRISA = 1;                      /* RA0 (AN0) is input */
        TRISB = 0;                      /* PORT B is output */
        TRISC = 0;                      /* RC0 is output */

        DA_Write = 1;                   /* Disable D/A converter */

        Initialize_AD();                /* Initialize A/D converter */
        Initialize_Timer();             /* Initialize timer interrupts */
        ei();                           /* Enable interrupts */
        for(;;);                        /* Wait for an interrupt */
}
```

Figure 10.20   (*Continued*)

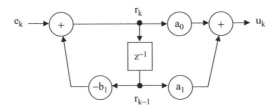

Figure 10.21 First-order module

Example 10.7

The circuit diagram of a digital control system is shown in Figure 10.18. A PIC16F877 micro-controller is to be used as the digital controller in this system. Assume that the set-point input s is to be hard-coded to the program, and the output y is analog and connected to A/D channel AN0 (bit 0 of port A) of the microcontroller. The microcontroller is assumed to operate with a crystal frequency of 4 MHz as shown in the figure. The output (port B) of the microcontroller is interfaced to a D/A converter which acts as a zero-order-hold and generates an analog output to drive the plant.

Assume that the digital controller to be implemented is in the form of a first-order module, and write a program in C to implement this controller. The controller parameters are assumed to be:

$$a_0 = 1, \quad a_1 = 0.8, \quad b_1 = 1.85,$$

i.e. the required controller transfer function is

$$D(z) = \frac{1 + 0.8z^{-1}}{1 + 1.85z^{-1}}.$$

Also assume that the required sampling interval is $T = 0.01$ s.

Solution

The implementation of a first-order module is very similar to a second-order module. The program to implement a first-order controller is shown in Figure 10.22. The operation of the program is very similar to the second-order program and is not described here.

10.5.3 Implementing Higher-Order Modules

Higher-order controllers can be implemented by cascading first-order and second-order modules. For example, a fourth-order controller can be implemented by cascading two second-order modules, as shown in Figure 10.23.

10.6 CHOICE OF SAMPLING INTERVAL

Whenever a digital control system is designed, a suitable sampling interval must be chosen. Choosing a large sampling time has destabilizing effects on the system. In addition, information loss occurs when large sampling times are selected. Also, the errors that occur when a continuous system is discretized increase as the sampling interval increases.

```
/*****************************************************************************

                         DIGITAL CONTROLLER
                         ==================

This program implements a first-order digital controller module on a
PIC16F877 (or equivalent) microcontroller.  The microcontroller operates with
a 4MHz crystal.  The analog input AN0 of the microcontroller is connected to
the output sensor of the plant (y).  The PORT B output of the microcontroller
is connected to an AD7302 type D/A converter.  The WR input of the controller
is controlled from port pin RC0 of the microcontroller.  The sampling
interval is 0.01s (10ms) and the timer interrupt service routine is used to
obtain the required sampling interval.

Program : First_Order_Module.C
Date : July 2005

******************************************************************************/
#include <pic.h>
#define DA_Write RC0

float DA_LSB,AD_LSB,a0,a1,b1,M1,M2,rk,rk_1,rk_2,ek,sk,yk,uk;
float y_high,y_low;

/* This function initializes the A/D converter so that analog data can be re-
ceived from channel AN0 of the microcontroller */
void Initialize_AD(void)
{
        ADCON1 = 0x8E;                  /* Configure AN0 */
        ADCON0 = 0x41;                  /* Select A/D clock and channel */
}


/* This function initilizes the timer TMR0 so that interrupts can be gener-
ated at 10ms intervals */
void Initialize_Timer(void)
{
        T0CS = 0;                       /* Select f/4 clock for TMR0 */
        PSA = 0;                        /* Select pre-scaler */
        PS0 = 1;                        /* Set pre-scaler to 64 */
        PS1 = 0;
        PS2 = 1;
        TMR0 = 100;                     /* Load TMR0=100 */
        T0IE = 1;                       /* Enable TMR0 interrupts */
        T0IF = 0;                       /* Clear TMR0 interrupt flag */
}


/* This function reads data from the A/D converter and stores in variable
yk */
void Read_AD_Input(void)
{
        ADCON0 = 0x45;                  /* Start A/D conversion */
        While((ADCON0 & 4) != 0);       /* Wait for conversion to complete */
        y_high = ADRESH;                /* High 2 bits of converted data */
        y_low = ADRESL;                 /* Low byte of converted data */
```

Figure 10.22   Complete program of the first-order controller (*Continued*)

```
        yk = 256.0*y_high + y_low;        /* Converted data in yk */
        yk = yk*AD_LSB;                   /* Sensor output in mV */
}


/* Interrupt Service Routine.  The program jumps here every 10 ms. */
void interrupt ISR(void)
{
        TMR0 = 100;                       /* Reload TMR0 */
        Read_AD_Input();                  /* Read A/D input */

        ek = sk - yk;                     /* Calculate error term */
        rk = ek + M1;
        yk = a0*rk + M2;                  /* Calculate output */
        uk = yk*DA_LSB;
        PORTB = uk;                       /* Send to PORT B */

        DA_Write = 0;                     /* Write to D/A converter */
        DA_Write = 1;

        rk_1 = rk;                        /* Update variables */
        M1 = -b1*rk_1;
        M2 = a1*rk_1;

        T0IF = 0;                         /* Re-enable timer interrupts */
}

/* Main Program.  The main program initializes the variables, A/D converter,
    D/A converter etc. and then waits in an endless loop for timer interrupts
    to Occur every 10 ms */

main(void)
{
        a0 = 1;        a1 = 0.8;      b1 = 1.85;
        M1 = 0;        M2 = 0;        rk = 0;    rk_1 = 0;
        sk = 1.0;
        AD_LSB = 5000.0/1024.0;
        DA_LSB = 256.0/5000.0;

        TRISA = 1;                        /* RA0 (AN0) is input */
        TRISB = 0;                        /* PORT B is output */
        TRISC = 0;                        /* RC0 is output */

        DA_Write = 1;                     /* Disable D/A converter */

        Initialize_AD();                  /* Initialize A/D converter */
        Initialize_Timer();               /* Initialize timer interrupts */
        ei();                             /* Enable interrupts */
        for(;;);                          /* Wait for an interrupt */
}
```

Figure 10.22   (*Continued*)

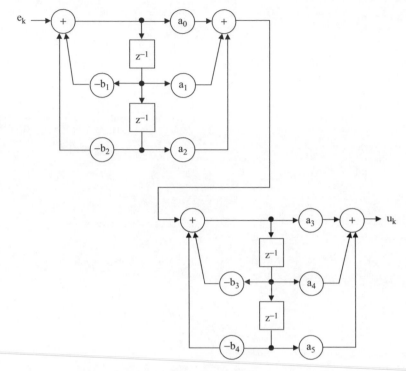

Figure 10.23 Implementing a fourth-order module

It may be thought that decreasing the sampling interval towards zero will make a discrete system converge towards an equivalent continuous system. However, in practice this is not the case since as the sampling interval is reduced, the change between the successive data values becomes less than the resolution of the system, leading to loss of information. In general, if a shorter sampling interval is to be used then the word length of the system should be increased so that the difference between adjacent samples can be resolved.

It has been found from practical applications in the process industry that a sampling interval of 1 s is generally short enough for most applications such as pressure control, temperature control and flow control. Systems with fast responses such as electromechanical systems (e.g. motors) require much shorter sampling intervals, usually of the order of milliseconds.

Various empirical rules have been suggested by many researchers for the selection of the sampling interval. These rules are based on practical experience and simulation results. Among them are the following

- If the plant has the dominant time constant T_p, then the sampling interval T for the closed-loop system should be selected such that $T < T_p/10$.

- Assuming that the process has a Ziegler–Nichols open-loop model

$$G(s) = \frac{e^{-sT_1}}{1 + sT_2},$$

then the sampling interval should be selected such that $T < T_1/4$.

- If the closed-loop system is required to have a settling time T_{ss} or a natural frequency of ω_n then choose the sampling interval T such that $T < T_{ss}/10$ and $\omega_s > 10\omega_n$, where ω_s is the sampling frequency, i.e. $\omega_s = 2\pi/T$.

10.7 EXERCISES

1. The transfer function of a digital controller is given by

$$D(z) = \frac{1 + 2z^{-1} + 3z^{-2}}{1 + 4z^{-1} + 5z^{-2}}.$$

    Draw the block diagram of the direct canonical realization of this controller.

2. Repeat Exercise 1 for a direct noncanocical controller realization.

3. Compare the realizations obtained in Exercises 1 and 2.

4. The transfer function of a digital controller is found to be

$$D(z) = \frac{1 + 2z^{-1} + 5z^{-2}}{1 + 3z^{-1} + 7z^{-2}}.$$

    Draw the block diagram of the direct noncanonical realization of this controller.

5. The transfer function of a digital controller is given by

$$D(z) = \frac{2(z + 2)(z + 3)}{z^2 + 0.4z + 0.03}.$$

    Use two first-order cascaded transfer functions to implement this controller.

6. The transfer function of a digital controller is given by

$$D(z) = \frac{(1 + 0.2z^{-1})(1 + 2z^{-1} + 4z^{-2})}{(1 + 0.3z^{-1})(1 + 0.2z^{-1} + 0.4z^{-2})}.$$

    Use a first-order and a second-order cascaded transfer function to implement this controller.

7. The transfer function of a digital controller is given by

$$D(z) = \frac{(1 + 2z^{-1})(1 + 3z^{-1})}{(1 + z^{-1})(1 + 5z^{-1})}.$$

    Realize this transfer function using first-order parallel transfer functions.

8. Draw the block diagram of the PID implementation using a parallel realization.

9. Draw the block diagram of the PID implementation using a direct canonical realization.

10. Describe how a given realization can be implemented on a microcontroller.

11. Draw a flow diagram to show how the PID algorithm can be implemented on a microcontroller. Write a program in C to implement this algorithm on a PIC microcontroller.

12. The transfer function of a digital controller is given by

$$D(z) = \frac{1 + 2z^{-1} + 5z^{-2}}{1 + 3z^{-1} + 4z^{-2}}.$$

Draw a flow diagram to show how this controller can be implemented on a microcomputer. Write a program in C to implement this algorithm on a PIC microcontroller.

13. Draw a flow diagram to show and explain how a second-order transfer function can be implemented on a PIC microcontroller using the C programming language.

14. Explain how second-order direct canonical functions can be cascaded to obtain higher-order transfer functions.

15. Explain how the sampling time can be selected in a first-order system.

16. Repeat Exercise 15 for a second-order system.

17. Describe the problems that may occur when very large or very small sampling times are selected.

18. Explain how the system stability is affected when the sampling time is increased.

FURTHER READING

[Åström and Hägglund, 1995]   Åström, K. and Hägglund, T. PID Controllers: Theory, design and tuning, 2nd edn., International Society for Measurement and Control. Research Triangle Park, NC, 1995.

[Cannon, 1967]   Cannon, R.H. Jr., Dynamics of Physical Systems. McGraw-Hill, New York, 1967.

[Cochin, 1980]   Cochin, I. Analysis and Design of Dynamic Systems. Harper & Row, New York, 1980.

[Crochiere and Oppenheim, 1975]   Crochiere, R.E. and Oppenheim, A.V. Analysis of linear digital networks. *Proc. IEEE*, **63**, April 1975, pp. 581–595.

[D'Souza, 1988]   D'Souza, A. Design of Control Systems. Prentice Hall, Englewood Cliffs, NJ, 1988.

[Fettweis, 1971]   Fettweis, A. Some principles of designing digital filters imitating classical filter structure. *IEEE Trans. Circuits Theory*, March 1971, pp. 314–316.

[Franklin and Powell, 1986]   Franklin, G.F. and Powell, J.D. Feedback Control of Dynamic Systems. Addison-Wesley, Reading, MA, 1986.

[Houpis and Lamont, 1992]   Houpis, C.H. and Lamont, B. Digital Control Systems: Theory, Hardware, Software, 2nd edn., McGraw-Hill, New York, 1992.

[Hwang, 1974]   Hwang, S.Y. An optimization of cascade fixed-point digital filters. *IEEE Trans. Circuits Syst. (Letters)*, **CAS-21**, January 1974, pp. 163–166.

[Jackson, 1970]   Jackson, L.B. Roundoff-noise analysis for fixed-point digital filters in cascade or parallel form. *IEEE Trans. Audio Electroacoust.* **AU-18**, June 1970, pp. 107–122.

[Katz, 1981]   Katz, P. Digital Control Using Microprocessors. Prentice Hall, Englewood Cliffs, NJ, 1981.

[Leigh, 1985]   Leigh, J.R. Applied Digital Control. Prentice Hall, Englewood Cliffs, NJ, 1985.

[Nagle and Nelson, 1981]   Nagle, H.T. Jr. and Nelson, V.P. Digital filter implementation on 16-bit micro-computers. *IEEE MICRO*, **1**, 1, February 1981, pp. 23–41.

[Ogata, 1990]   Ogata, K. Modern Control Engineering, 2nd edn., Prentice Hall, Englewood Cliffs, NJ, 1990.

[Phillips and Nagle, 1990]   Phillips, C.L. and Nagle, H.T. Jr. Digital Control Systems: Analysis and Design, 2nd edn., Prentice Hall, Englewood Cliffs, NJ, 1990.

11

Liquid Level Digital Control System: A Case Study

In this chapter we shall look at the design of a digital controller for a control system, namely a liquid level control system.

Liquid level control systems are commonly used in many process control applications to control, for example, the level of liquid in a tank. Figure 11.1 shows a typical liquid level control system. Liquid enters the tank using a pump, and after some processing within the tank the liquid leaves from the bottom of the tank. The requirement in this system is to control the rate of liquid delivered by the pump so that the level of liquid within the tank is at the desired point.

In this chapter the system will be identified from a simple step response analysis. A constant voltage will be applied to the pump so that a constant rate of liquid can be pumped to the tank. The height of the liquid inside the tank will then be measured and plotted. A simple model of the system can then be derived from this response curve. After obtaining a model of the system, a suitable controller will be designed to control the level of the liquid inside the tank.

11.1 THE SYSTEM SCHEMATIC

The schematic of the liquid level control system used in this case study is shown in Figure 11.2. The system consists of a water tank, a water pump, a liquid level sensor, a microcontroller, a D/A converter and a power amplifier.

Water tank. This is the tank where the level of the liquid inside is to be controlled. Water is pumped to the tank from above and a level sensor measures the height of the water inside the tank. The microcontroller controls the pump so that the liquid is at the required level. The tank used in this case study is a plastic container with measurements 12 cm × 10 cm × 10 cm.

Water pump. The pump is a small 12 V water pump drawing about 3 A when operating at the full-scale voltage. Figure 11.3 shows the pump.

Level sensor. A rotary potentiometer type level sensor is used in this project. The sensor consists of a floating arm connected to the sliding arm of a rotary potentiometer. The level of the floating arm, and hence the resistance, changes as the liquid level inside the tank is changed. A voltage is applied across the potentiometer and the change of voltage is measured across the arm of the potentiometer. The resistance changes from 430Ω when the floating arm is at the bottom (i.e. there is no liquid inside the tank) to 40 Ω when the arm is at the top. The level sensor is shown in Figure 11.4.

Microcontroller Based Applied Digital Control   D. Ibrahim
© 2006 John Wiley & Sons, Ltd

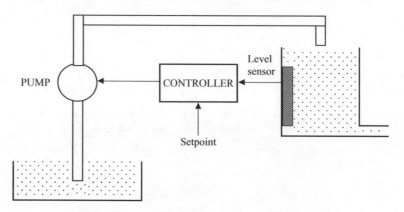

Figure 11.1 A typical liquid level control system

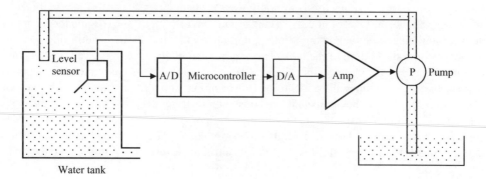

Figure 11.2 Schematic of the system

Microcontroller. A PIC16F877 type microcontroller is used in this project as the digital controller. In general, any other type of microcontroller with a built-in A/D converter can be used. The PIC16F877 incorporates an 8-channel, 10-bit A/D converter.

D/A converter. An 8-bit AD7302 type D/A converter is used in this project. In general, any other type of D/A converter can be used with similar specifications.

Power amplifier. The output power of the D/A converter is limited to a few hundred milliwatts, which is not enough to drive the pump. An LM675 type power amplifier is used to increase the power output of the D/A converter and drive the pump. The LM675 can provide around 30 W of power.

11.2 SYSTEM MODEL

The system is basically a first-order system. The tank acts as a fluid capacitor where fluid enters and leaves the tank. According to mass balance,

$$Q_{in} = Q + Q_{out}, \tag{11.1}$$

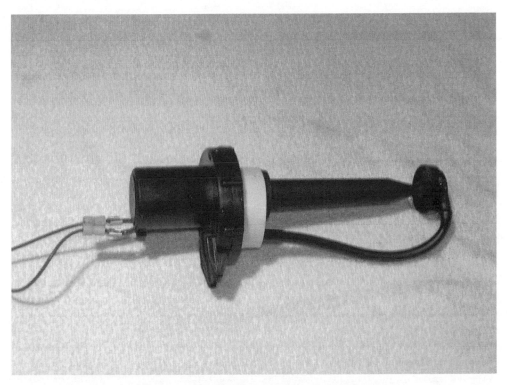

Figure 11.3   The pump used in the project

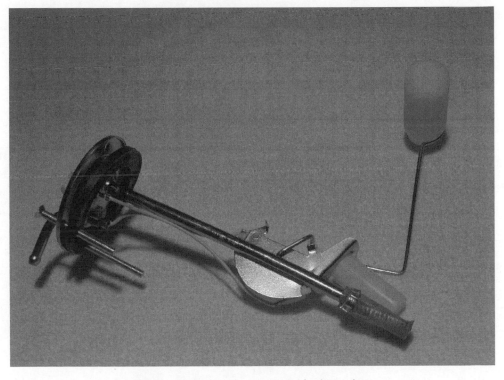

Figure 11.4   Level sensor used in the project

where Q_{in} is the flow rate of water into the tank, Q the rate of water storage in the tank, and Q_{out} the flow rate of water out of the tank. If A is the cross-sectional area of the tank, and h the height of water inside the tank, (11.1) can be written as

$$Q_{in} = A\frac{dh}{dt} + Q_{out}. \tag{11.2}$$

The flow rate of water out of the tank depends on the discharge coefficient of the tank, the height of the liquid inside the tank, the gravitational constant, and the area of the tank outlet, i.e.

$$Q_{out} = C_d a\sqrt{2gh}, \tag{11.3}$$

where C_d is the discharge coefficient of the tank outlet, a the are of the tank outlet, and g the gravitational constant (9.8 m/s²).

From (11.2) and (11.3) we obtain

$$Q_{in} = A\frac{dh}{dt} + C_d a\sqrt{2gh}. \tag{11.4}$$

Equation (11.4) shows a nonlinear relationship between the flow rate and the height of the water inside the tank. We can linearize this equation for small perturbations about an operating point.

When the input flow rate Q_{in} is a constant, the flow rate through the orifice reaches a steady-state value $Q_{out} = Q_0$, and the height of the water reaches the constant value h_0, where

$$Q_0 = C_d a\sqrt{2gh_0}. \tag{11.5}$$

If we now consider a small perturbation in input flow rate around the steady-state value, we obtain

$$\delta Q_{in} = Q_{in} - Q_0 \tag{11.6}$$

and, as a result, the fluid level will be perturbed around the steady-state value by

$$\delta h = h - h_0. \tag{11.7}$$

Now, substituting (11.6) and (11.7) into (11.4) we obtain

$$A\frac{d\delta h}{dt} + C_d a\sqrt{2(\delta h + h_0)} = \delta Q_{in} + Q_0. \tag{11.8}$$

Equation (11.8) can be linearized by using the Taylor series and taking the first term. From Taylor series,

$$f(x) = f(x_0) + \frac{df}{dx}\bigg|_{x=x_0} \frac{(x - x_0)}{1!} + \frac{d^2 f}{dx^2}\bigg|_{x=x_0} \frac{(x - x_0)^2}{2!} + \cdots. \tag{11.9}$$

Taking only the first term,

$$f(x) - f(x_0) \approx \frac{df}{dx}\bigg|_{x=x_0} (x - x_0) \tag{11.10}$$

or

$$\delta f(x) \approx \frac{df}{dx}\bigg|_{x=x_0} \delta x. \tag{11.11}$$

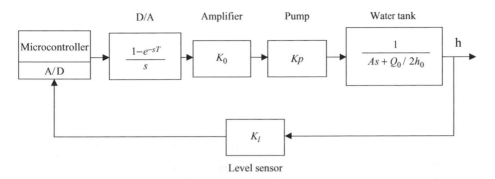

Figure 11.5 Block diagram of the system

Linearizing (11.8) using (11.11), we obtain

$$A\frac{d\delta h}{dt} + \frac{Q_0}{2h_0}\delta h = \delta Q_{\text{in}}.$$ (11.12)

Taking the Laplace transform of (11.12), we obtain the transfer function of the tank for small perturbations about the steady-state value as a first-order system:

$$\frac{h(s)}{Q_{\text{in}}(s)} = \frac{1}{As + Q_0/2h_0}.$$ (11.13)

The pump, level sensor, and the power amplifier are simple units with proportional gains and no system dynamics. The input–output relations of these units can be written as follows: for the pump,

$$Q_p = K_p V_p;$$

for the level sensor,

$$V_l = K_l h;$$

and for the power amplifier,

$$V_0 = K_0 V_i.$$

Here Q_p is the pump flow rate, V_p the voltage applied to the pump, V_l the level sensor output voltage, V_0 the output voltage of the power amplifier, and V_i the input voltage of the power amplifier; K_p, K_l, K_0 are constants.

The block diagram of the level control system is shown in Figure 11.5.

11.3 IDENTIFICATION OF THE SYSTEM

The system was identified by carrying out a simple step response test. Figure 11.6 shows the hardware set-up for the step response test. The port B output of the microcontroller is connected to data inputs of the D/A converter, and the converter is controlled from pin RC0 of the microcontroller. The output of the D/A converter is connected to the LM675 power amplifier which drives the pump. The value of the step was chosen as 200, which corresponds to a D/A voltage of $5000 \times 200/256 = 3.9$ V. The height of the water inside the tank (output of the level

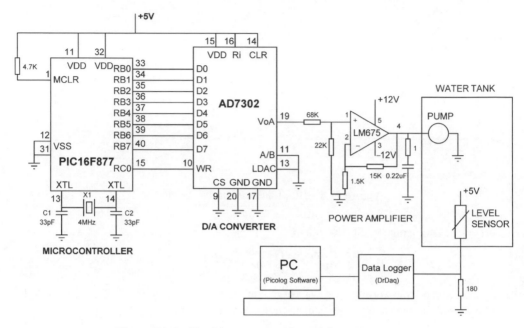

Figure 11.6 Hardware set-up to record the step response

sensor) was recorded in real time using a *DrDaq* type data logger unit and the *Picolog* software. Both of these products are manufactured by PICO Technology. *DrDaq* is a small electronic card which is plugged into the parallel port of a PC. The card is equipped with sensors to measure physical quantities such as the intensity, sound level, voltage, humidity, and temperature. *Picolog* runs on a PC and can be used to record the measurements of the *DrDaq* card in real time. The software includes a graphical option which enables the measurements to be plotted.

The microcontroller program to send a step signal to the D/A converter is shown in Figure 11.7. At the beginning of the program the input–output ports are configured and then a step signal (200) is sent to port B. The D/A converter is then enabled by clearing its WR input. After writing data to the D/A converter it is disabled so that its output does not accidentally change. The program then waits in an endless loop.

Figure 11.8 shows the step response of the system, which is the response of a typical first-order system. It will be seen that the response contains noise. Also, since the *DrDaq* data logger is 8-bit, its resolution is about 19.5 mV with a reference input of 5V, and this causes the step discontinuities shown in the response (the steps can be eliminated either by using a data logger with a higher resolution, or by amplifying the output of the level sensor). The figure clearly shows that in practice the response of a system is not always a perfect textbook signal.

A smooth curve is drawn through the response by taking the midpoints of the steps, as shown in Figure 11.9.

11.4 DESIGNING A CONTROLLER

The circuit diagram of the closed-loop system is shown in Figure 11.10. The loop is closed by connecting the output of the level sensor to the analog input AN0 of the microcontroller.

```
/*----------------------------------------------------------------

                        STEP RESPONSE TEST
                        ==================

This program sends a STEP input to the D/A converter.  The value
of the input is set to 200, which corresponds to a voltage of
5 × 200/256V.  The output of the D/A is connected to a power
amplifier which has an overall gain of G = 2.5.  Thus, the step
voltage applied to the pump is 5 × (200/256) × 2.5 = 9.76V.

The hardware consists of a PIC16F877 microcontroller, where PORT B
is conencted to an AD7302 type D/A converter.  The output of the D/A
is connected to an LM675 type DC power amplifier.


File: STEP.C
Date: July 2005

---------------------------------------------------------------- */


#include <pic.h>

#define AD7302_WR RC0

/* Start of main program */
main(void)
{
TRISB = 0;                           /* PORTB is output */
TRISC = 0;                           /* RC0 is output */
AD7302_WR = 1;                       /* Disable D/A */

/* Send a STEP input to the D/A */
PORTB = 200;                         /* Send the STEP */
AD7302_WR = 0;                       /* Enable D/A */
AD7302_WR = 1;                       /* Disable D/A */

wait: goto wait;                     /* Wait here forever */

}
```

Figure 11.7 Microcontroller program to send a step to D/A

A controller algorithm was then implemented in the microcontroller to control the level of the water in the tank.

One of the requirements in this case study is zero steady-state error, which can be achieved by having an integral type controller. In this case study a Ziegler–Nichols PI controller was designed.

The system model can be derived from the step response. As shown in Figure 11.11, the Ziegler–Nichols system model parameters are given by $T_1 = 31$ s, $T_D = 2$ s and

$$K = \frac{2345 - 2150}{200 \times 5000/256} = 0.05.$$

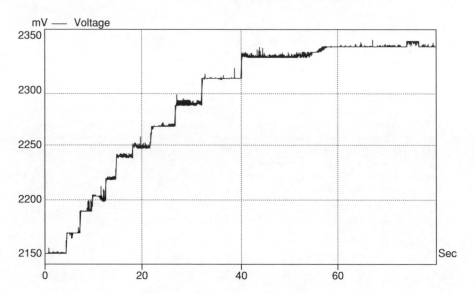

Figure 11.8 System step response

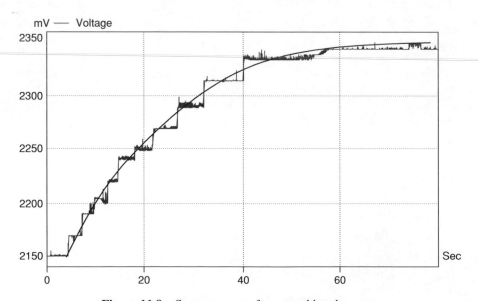

Figure 11.9 Step response after smoothing the curve

Notice that the output of the microcontroller was set to 200, which corresponds to $200 \times 5000/256 = 3906$ mV, and this was the voltage applied to the power. We then obtain the following transfer function:

$$G(s) = \frac{0.05e^{-2s}}{1 + 31s} \tag{11.14}$$

The time constant of the system is 31 s. It was shown in Section 10.6 that the sampling time should be chosen to be less than one-tenth of the system time constant, i.e. $T < 3.1$ s. In this case study, the sampling time is chosen to be 100 ms, i.e. $T = 0.1$ s.

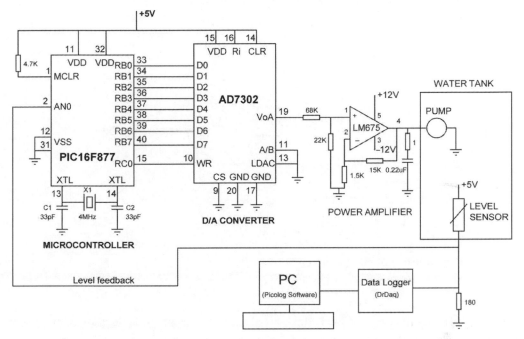

Figure 11.10   Circuit diagram of the closed-loop system

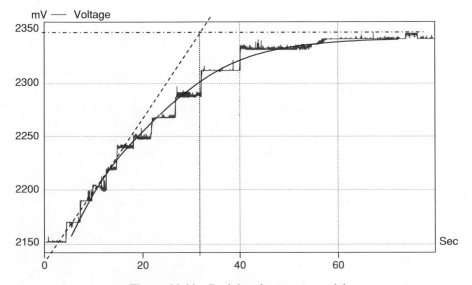

Figure 11.11   Deriving the system model

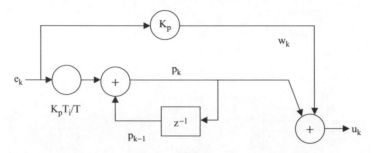

Figure 11.12   Realization of the controller

The coefficients of a Ziegler–Nichols PI controller were given in Chapter 10:

$$K_p = \frac{0.9T_1}{KT_D} \quad \text{and} \quad T_i = 3.3T_D.$$

Thus, the PI parameters of our system are:

$$K_p = \frac{0.9 \times 31}{0.05 \times 2} = 279 \quad \text{and} \quad T_i = 3.3 \times 2 = 6.6.$$

A parallel PI controller was realized in this case study. The controller is in the form of (10.27), with the derivative term set to zero, i.e.

$$D(z) = K_p + \frac{K_p T}{T_i(1 - z^{-1})}.$$

The realization of the controller as a parallel structure is shown in Figure 11.12.

The controller software is shown in Figure 11.13. The PI algorithm has been implemented as a parallel structure. At the beginning of the program the controller parameters are defined. The program consists of the functions *Initialize_AD*, *Initialize_Timer*, *Read_AD_Input*, and the interrupt service routine (ISR). The A/D converter is initialized to receive analog data from channel AN0. The *Read_AD_Input* function reads a sample from the A/D converter and stores it in variable y_k. The timer is initialized to interrupt every 10 ms. At the beginning of the ISR routine, the PI algorithm is implemented after every 10th interrupt, i.e. every 100 ms. This ensures that the controller sampling time is 100 ms. The ISR routine reads the output of the level sensor and converts it to digital. Then the PI controller algorithm is implemented. Notice that in the algorithm the input of the D/A converter is limited to full scale, i.e. 255. After sending an output to D/A, the ISR routine re-enables the timer interrupts and the program waits for the occurrence of the next interrupt.

The step response of the closed-loop system is shown in Figure 11.14. Here, the reference input was set to 2280. Clearly the system response, although noisy, reaches the set-point with no steady-state error, as desired.

11.5   CONCLUSIONS

Although the case study given in this chapter is simple, it illustrates the basic principles of designing a digital controller, from the identification of the system to the implementation of a suitable controller algorithm on a microcontroller. Here the classical Ziegler–Nichols PI type

```
/***********************************************************************

                    ZIEGLER-NICHOLS PI CONTROLLER
                    =============================

This program implements a first-order digital controller module on a
PIC16F877 (or equivalent) microcontroller.  The microcontroller operates
with a 4MHz crystal. The analog input AN0 of the microcontroller is connected
to the output sensor of the plant (y).  The PORT B output of the
microcontroller is connected to an AD7302 type D/A converter.  The WR input
of the controller is controlled from port pin RC0 of the microcontroller.
The sampling interval is 0.1s (100ms) and the timer interrupt service
routine is used to obtain the required sampling interval.

Program : ZIEGLER.C
Date : July 2005

***********************************************************************/

#include <pic.h>

#define DA_Write RC0

volatile unsigned int time_now = 0;

float AD_LSB, DA_LSB, wk,Kp, T, Ti, b, pk, pk_1, ek, sk, yk, y_high, y_low;

unsigned char uk;

/* This function initializes the A/D converter so that analog data can be
received from channel AN0 of the microcontroller */

void Initialize_AD(void)

{
ADCON1 = 0x8E;               /* Configure AN0 for +5V reference */
ADCON0 = 0x41;               /* Select A/D converter clock */
}

/* This function initilizes the timer TMR0 so that interrupts can be
generated at 10ms intervals */

void Initialize_Timer(void)

{
T0CS = 0;                    /* Select f/4 clock for the TMR0 */
PSA = 0;                     /* Select pre-scaler */
PS0 = 1;                     /* Set pre-scaler to 64 */
PS1 = 0;                     /* PS2,PS1,PS0 = 101 */
PS2 = 1;
TMR0 = 100;                  /* Load TMR0 = 100 */
T0IE = 1;                    /* Enable TMR0 interrupts */
T0IF = 0;                    /* Clear TMR0 interrupt flag */
}
```

Figure 11.13   Software of the controller (*Continued*)

```
/* This function reads data from the A/D converter and stores it in
variable yk */
void Read_AD_Input(void)

{
ADCON0 = 0x45;                          /* Start A/D conversion */
while((ADCON0 & 4) != 0);               /* Wait until convertion completes */
y_high = ADRESH;                        /* High 2 bytes of converted data */
y_low = ADRESL;                         /* Low byte of converted data */
yk = 256.0*y_high + y_low;              /* Converted data in yk */
yk = yk*AD_LSB;                         /* Sensor output in mV */
}


/* Interrupt Service Routine.  The program jumps here every 10 ms. */
void interrupt ISR(void)

{
TMR0 = 100;                             /* Reload TMR0 = 100 */
/* Check if 100ms has elapsed. The basic timer rate is 10ms and we
   count 10 interrupts before implementing the controller algorithm */

time_now++;
if(time_now == 10)
{
time_now = 0;
Read_AD_Input();                        /* Read A/D input */

ek = sk - yk;                           /* Calculate error term */
pk = b*ek + pk_1;
wk = Kp*ek;                             /* Calculate output */
yk = wk + pk;
yk = yk*DA_LSB;
if(yk > 255)
uk=255;
else
uk=(unsigned char)yk;
PORTB = uk;

DA_Write = 0;                           /* Write to D/A converter */
DA_Write = 1;

pk_1 = pk;
}
T0IF = 0;                               /* Re-enable timer interrupts */
}

/* Main Program.  The main program initializes the variables, A/D converter,
   D/A converter etc. and then waits in an endless loop for timer inter-
rupts to
   Occur every 100 ms */

main(void)
{
Kp = 279.0; T = 0.1;      Ti = 6.6;
b = Kp*T/Ti;
pk = 0.0;    pk_1 = 0.0;
```

Figure 11.13   (*Continued*)

```
AD_LSB = 5000.0/1024.0;
DA_LSB = 256.0/5000.0;

TRISA = 1;
TRISB = 0;                      /* PORTB is output */
TRISC = 0;                      /* RC0 is output */

sk = 2280.0;                    /* Setpoint input */

DA_Write = 1;                   /* Disable D/A converter */

Initialize_AD();                /* Initialize A/D converter */
Initialize_Timer();             /* Initialize timer interrupts */
ei();                           /* Enable interrupts */

for(;;);                        /* Wait for an interrupt */
}
```

Figure 11.13   (*Continued*)

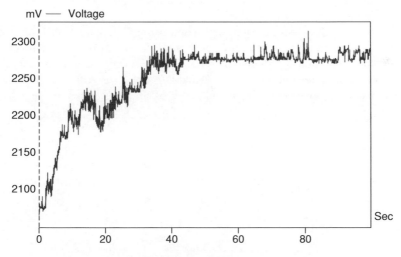

Figure 11.14   Step response of the closed-loop system

controller has been implemented. The Ziegler–Nichols tuning method is based on a quarter-amplitude decay ratio, which means that the amplitude of an oscillation is reduced by a factor of 4 over a whole period. This corresponds to a damping ratio of about 0.2, which gives rise to a large overshoot. Ziegler–Nichols tuning method has been very popular and most manufacturers of controllers have used these rules with some modifications. The biggest advantage of the Ziegler–Nichols tuning method is its simplicity.

In practice one has to consider various implementation-related problems, such as:

- integral wind-up of the PID controller;

- the derivative kick of the PID controller;

- quantization errors produced as a result of the finite word-length data storage and finite word-length arithmetic;

- errors produced as a result of the finite word length of the A/D and the D/A converters.

Quantization errors are an important topic in the design of digital controllers using microcontrollers and programming languages with fixed-point arithmetic. Numbers have to be approximated in order to fit the word length of the digital computer used. Approximations occur at the A/D conversion stage. The A/D converter represents +5 V with 8, 12 or 16 bits. The resolution is then 19.53 mV for an 8-bit converter, 4.88 mV for a 12-bit converter, and 0.076 mV for a 16-bit converter. If, for example, we are using an 8-bit A/D converter then a signal change less than 19.53 mV will not be recognized by the system.

Approximations also occur at the D/A conversion stage where the output of the digital algorithm has to be converted into an analog signal to drive the plant. The finite resolution of the D/A converter introduces errors into the algorithm.

Quantization errors occur after the mathematical operations inside the microcontroller, especially after a multiplication where the result must be truncated or rounded before being stored.

Controller parameters are not integers, and quantization errors occur when these constants are introduced into the controller algorithm.

When floating-point arithmetic is used (e.g. when using C language) in 8-bit microcontrollers, numbers can be stored with very high accuracy. The effects of the quantization errors due to mathematical operations inside the microcontroller are very small in such applications. The errors due to the finite resolution of the A/D and D/A converters can still introduce errors into the system and cause unexpected behaviour.

Appendix A
Table of z-Transforms

$f(kT)$	$F(z)$
$\delta(t)$	1
1	$\dfrac{z}{z-1}$
kT	$\dfrac{Tz}{(z-1)^2}$
$(kT)^2$	$\dfrac{T^2z(z+1)}{2(z-1)^3}$
$(kT)^3$	$\dfrac{T^3z(z^2+4z+1)}{(z-1)^4}$
e^{-akT}	$\dfrac{z}{z-e^{-aT}}$
kTe^{-akT}	$\dfrac{Tze^{-aT}}{(z-e^{-aT})^2}$
a^k	$\dfrac{z}{z-a}$
$1-e^{-akT}$	$\dfrac{z(1-e^{-aT})}{(z-1)(z-e^{-aT})}$
$\sin akT$	$\dfrac{z\sin aT}{z^2-2z\cos aT+1}$
$\cos akT$	$\dfrac{z(z-\cos aT)}{z^2-2z\cos aT+1}$
$e^{-akT}\sin bkT$	$\dfrac{e^{-aT}z\sin bT}{z^2-2e^{-aT}z\cos bT+e^{-2aT}}$
$e^{-akT}\cos bkT$	$\dfrac{z^2-e^{-aT}z\cos bT}{z^2-2e^{-aT}z\cos bT+e^{-2aT}}$

Laplace transform	Corresponding z-transform
$\dfrac{1}{s}$	$\dfrac{z}{z-1}$
$\dfrac{1}{s^2}$	$\dfrac{Tz}{(z-1)^2}$
$\dfrac{1}{s^3}$	$\dfrac{T^2z(z+1)}{2(z-1)^3}$
$\dfrac{1}{s+a}$	$\dfrac{z}{z-e^{-aT}}$
$\dfrac{1}{(s+a)^2}$	$\dfrac{Tze^{-aT}}{(z-e^{-aT})^2}$
$\dfrac{a}{s(s+a)}$	$\dfrac{z(1-e^{-aT})}{(z-1)(z-e^{-aT})}$
$\dfrac{b-a}{(s+a)(s+b)}$	$\dfrac{z(e^{-aT}-e^{-bT})}{(z-e^{-aT})(z-e^{-bT})}$
$\dfrac{(b-a)s}{(s+a)(s+b)}$	$\dfrac{(b-a)z^2-(be^{-aT}-ae^{-bT})z}{(z-e^{-aT})(z-e^{-bT})}$
$\dfrac{a}{s^2+a^2}$	$\dfrac{z\sin aT}{z^2-2z\cos aT+1}$
$\dfrac{s}{s^2+a^2}$	$\dfrac{z^2-z\cos aT}{z^2-2z\cos aT+1}$
$\dfrac{s}{(s+a)^2}$	$\dfrac{z[z-e^{-aT}(1+aT)]}{(z-e^{-aT})^2}$

Appendix B
MATLAB Tutorial

This tutorial is an introduction to MATLAB. MATLAB is an interactive environment for scientific and engineering calculations, design, simulation and visualization. The aim of this tutorial is to enable students and control engineers to learn to use MATLAB in control engineering applications. The tutorial provides a brief introduction to the use of MATLAB with examples, and then describes the *Control System Toolbox* with examples. With the aid of this toolbox, for example, the control engineer can draw the Bode diagram, root locus or time response of a system in a few seconds and analyse and design a control system in a very short time.

B.1 MATLAB OPERATIONS

A variable in MATLAB can be a scalar, a complex number, a vector, or a matrix. A variable name can be up to 31 characters long and must start with a letter. It can contain letters, numbers, and underscore characters.

If a data entry, a statement, or any command is not terminated by a semicolon, the result of the statement is always displayed.

An integer number can be assigned to a variable name as follows:

```
>> w = 5;
>> p = -3;
```

A real number is entered by specifying a decimal point, or the exponent:

```
>> q = 2.35;
>> a = 5.2e-3;
```

When entering a complex number, the real part is entered first, followed by the imaginary part:

```
>> p = 2 + 4*i;
>> q = 12*exp (i*2);
```

Microcontroller Based Applied Digital Control   D. Ibrahim
© 2006 John Wiley & Sons, Ltd

A row vector is entered by optionally separating the elements with commas. For example, the vector

$$p = [2 \quad 4 \quad 6]$$

is entered as

```
>> p = [2, 4, 6];
```

or

```
>> p = [2 4 6];
```

Similarly, a column vector is entered by separating the elements with semicolons. For example, the vector

$$q = \begin{bmatrix} 3 \\ 6 \\ 9 \end{bmatrix}$$

is entered as

```
>> q = [3; 6; 9];
```

A vector can be transposed by using the character ''. For example, the vector q above can be transposed as

```
>> a = [3; 6; 9]';
```

to give

```
a = [3 6 9].
```

Matrices are entered similarly to vectors: the elements of each row can be entered by separating them with commas; the rows are then separated by the semicolon character. For example, the 3×3 matrix A given by

$$A = \begin{bmatrix} 2 & 1 & 3 \\ 4 & 0 & 6 \\ 5 & 8 & 7 \end{bmatrix}$$

is entered as

```
>> A = [2, 1, 3; 4, 0, 6; 5, 8, 7];
```

or

```
>> A = [2 1 3; 4 0 6; 5 8 7];
```

Special vectors and matrices. MATLAB allows special vectors and matrices to be declared. Some examples are given below:

```
A = []          generates a null matrix
A = ones(n,m)   generates an n × m matrix of ones

A = eye(n)      generates an n × n identity matrix
A = zeros(n,m)  generates an n × m matrix of zeros
```

Some examples are given below:

```
>> A = ones(3,5)
```

gives

```
A =
    1  1  1  1  1
    1  1  1  1  1
    1  1  1  1  1
>> B = zeros(2,3)
```

gives

```
B =
    0  0  0
    0  0  0
```

and

```
>> C = eye(3,3)
```

gives

```
C =
    1  0  0
    0  1  0
    0  0  1
```

A particular element of a matrix can be assigned by specifying the row and the column numbers:

```
>> A(2,3) = 8;
```

places the number 8 in the second row, third column.

Matrix elements can be accessed by specifying the row and the column numbers:

```
>> C = A(2,1);
```

assigns the value in the second row, first column of A to C.

Vectors can be created by specifying the initial value, final value and increment. For example,

```
>> T = 0:1:10;
```

creates a row vector called T with the elements:

```
T = [0  1  2  3  4  5  6  7  8  9  10].
```

If the increment is 1 it can be omitted. Thus the above vector can also be created with the statement

```
>> T = 0:10;
```

The size of a matrix can be found by using the *size* statement:

```
>> [m,n] = size(C)

m =
    4
n =
    3
```

Arithmetic operators. MATLAB utilizes the following arithmetic operators:

```
+          addition
-          subtraction
*          multiplication
/          division
^          power operator
'          transpose
```

If x is a vector, its multiplication with a scalar multiplies all elements of the vector. For example,

```
>> x = [1,  3,  5];
>> y = 2*x
y =
    2  6  10
```

Similarly, if A is a matrix, its multiplication with a scalar multiplies all elements of the matrix:

```
>> A = [1  3;  5  7];
>> B = 2*A

B =
    2   6
   10  14
```

Two matrices can be multiplied to produce another matrix. For example, if

$$A = \begin{bmatrix} 1 & 3 \\ 2 & 4 \end{bmatrix} \quad \text{and} \quad B = \begin{bmatrix} 2 & 4 \\ 5 & 2 \end{bmatrix}$$

then

```
>> A = [1  3;  2  4];
>> B = [2  4;  5  2];
>> C = A*B

C =
   17  10
   24  16
```

Array operations perform arithmetic operations in an element-by-element manner. An array operation is indicated by proceeding the operator by a period (.). For example, if $a = \begin{bmatrix} 1 & 3 & 4 \end{bmatrix}$ and $b = \begin{bmatrix} 2 & 3 & 5 \end{bmatrix}$ then

```
>> a = [1  3  4];
>> b = [2  3  5];
>> c = a.*b

c =
    2  9  20
```

Predefined functions. There are a number of predefined functions that can be used in statements. Some commonly used functions are:

abs	absolute value
sqrt	square root
real	real part
imag	imaginary part
rem	remainder
sin	sine
cos	cosine
asin	arcsine
acos	arccosine
tan	tangent
atan	arctangent
exp	exponential base e
log	natural logarithm
log10	log base 10

For example,

```
>> a = sqrt(16)
a = 4
>> a = sqrt(-4)
a = 0 + 2.0000i
```

Polynomials. A polynomial is defined by using a vector containing the coefficients of the polynomial. For example, the polynomial

$$F(x) = 3x^4 - 5x^3 + x^2 - 3x + 1$$

is defined as

```
p = [3  -5  1  -3  1].
```

It is important that all powers of the polynomial must be specified. The coefficients of the missing powers must be specified as zero.

The following operations can be performed on a polynomial:

roots(p)	find the roots of the polynomial
polyval(p,x)	evaluate the polynomial p at the value of x
deconv(p1,p2)	compute the quotient of p_1 divided by p_2
conv(p1,p2)	compute the product of polynomials p_1 and p_2
poly(r)	compute the polynomial from the vector of $r$oots
poly2str(p,'s')	display the polynomial as an equation in s

For example, consider the polynomial P_1, where,

$$P_1 = 6x^4 - 2x^3 + 5x^2 - 2x + 1.$$

The roots of $P_1 = 0$ are found as follows:

```
>> P1 = [6  -2  5  -2  1]'
>> r = roots(P1)


r =
    -0.1026 + 0.8355i
    -0.1026 - 0.8355i
     0.2692 + 0.4034i
     0.2692 - 0.4034i
```

The polynomial has four complex roots.

The value of the polynomial at $x = 1.2$ is 14.7856 and can be found as follows:

```
>> polyval(P1, 1.2)

ans =

    14.7856
```

The polynomial P_1 can be expressed as an equation in s as:

```
>> poly2str(P1,'s')

ans =
    6 s^4 - 2 s^3 + 5s^2 - 2 s + 1
```

Now consider another polynomial

$$P_2 = 2x^4 - x^3 + 2x^2 - x + 3.$$

The product of the polynomials P1 and P2 can be found as follows:

```
>> P2 = [2  -1  2  -1  3];
>> P3 = conv(P1,P2)

P3 =
    12  -10  24  -19  34  -16  19  -7  3
```

or

```
>> P3 = poly2str(conv(P1,P2),'x')

P3 =
    12 x^8 - 10 x^7 + 24 x^6 - 19 x^5 + 34 x^4 - 16 x^3 + 19 x^2 -
    7 x + 3
```

Finally, the polynomial whose roots are 2 and 5 can be found as follows:

```
>> poly([2 5])

ans =

    1  -7  10
```

or

```
>> poly2str(poly([2   5]),'x')
```

ans =

    x^2 - 7x + 10

Thus, the equation of the required polynomial is

$$F(x) = x^2 - 7x + 10.$$

B.2   CONTROL SYSTEM TOOLBOX

The Control System Toolbox is a collection of algorithms and uses MATLAB functions to provide specilized functions in control engineering. In this section we will briefly look at some of the important functions of the Control System Toolbox for both continuous-time and discrete-time systems.

B.2.1   Continuous-Time Systems

Consider a single-input, single-output continuous-time system with the open-loop transfer function

$$\frac{Y(s)}{U(s)} = \frac{3}{s^2 + 3s + 9}.$$

Transfer function. The transfer function of the system can be defined in terms of the numerator and the denominator polynomial:

```
>> num = [0   0   3];
>> den = [1   3   9];
```

The transfer function is given by:

```
>> G = tf(num,den)
```

Transfer function:

```
        3
   -----------
   s^2 + 3s + 9
```

Step response. The step response is given by

```
>> step(num,den)
```

which produces the plot shown in Figure B.1.

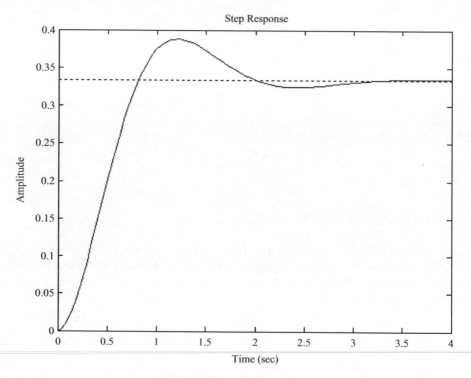

Figure B.1   Step response

The steady-state value of the step response is obtained as follows:

```
>> ys = decgain(num,den)

ys =

    0.3333
```

Impulse response. The impulse response is given by

```
>> impulse(num,den)
```

which produces the plot shown in Figure B.2.

Bode diagram. The Bode diagram can be obtained by writing:

```
>> impulse(num,den);
>> bode(num,den);
>> grid
```

Notice that the *grid* command produces a grid in the display. The Bode diagram of the system is shown in Figure B.3.

Nyquist diagram. The Nyquist diagram can be obtained from

```
>> nyquist(num,den)
```

The Nyquist diagram of the system is shown in Figure B.4.

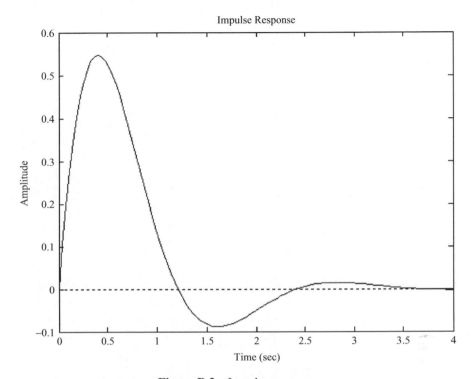

Figure B.2   Impulse response

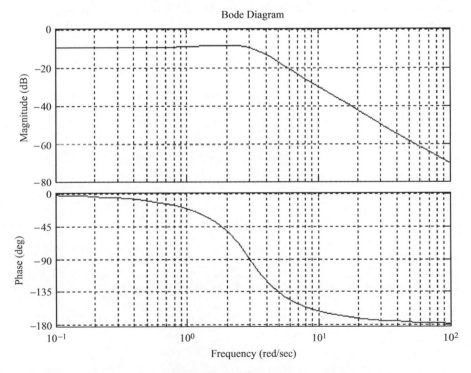

Figure B.3   Bode diagram

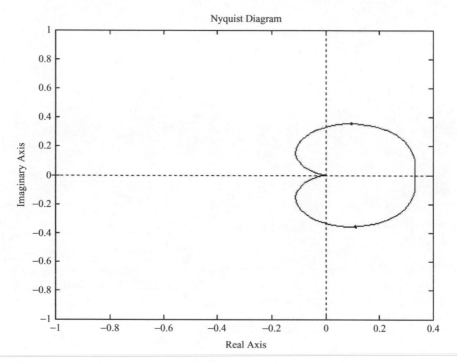

Figure B.4 Nyquist diagram

Nichols diagram. The Nichols diagram can be obtained by writing

```
>> nichols(num,den);
>> grid
```

Figure B.5 shows the Nichols diagram obtained.

Root locus. The root locus diagram of the system is given by

```
>> rlocus(num,den)
```

Figure B.6 shows the graph obtained from this command.

To customize the plot for a specific range of K, say for K ranging from 0 to 5 in steps of 0.5, we can write

```
>> K = 0:0.5:5;
>> r = rlocus(num,den,K);
>> plot(r,'x')
```

Figure B.7 shows the graph obtained, where the character x is plotted at the roots of the system as K is varied.

Zeros and poles. The zero and pole locations can be found from

```
>> [z,p,k] = tf2zp(num,den)

z =
     Empty matrix: 0-by-1
```

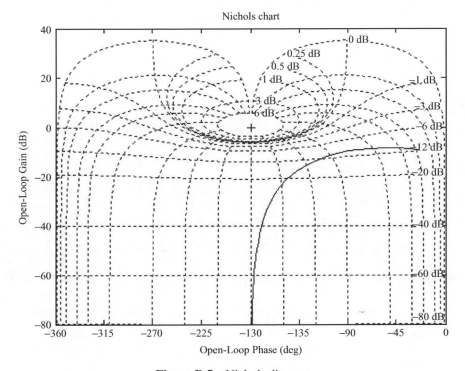

Figure B.5   Nichols diagram

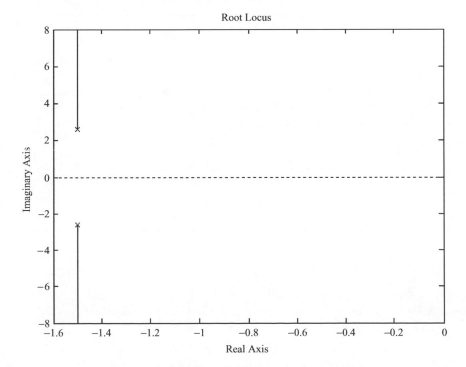

Figure B.6   Root locus diagram

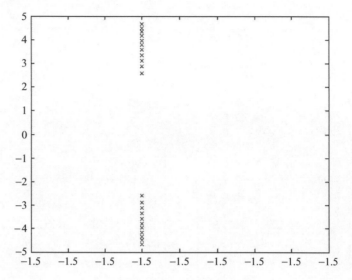

Figure B.7   Root locus diagram

```
p =
    -1.5000 + 2.5981i
    -1.5000 - 2.5981i

k =
    3
```

where z is the zeros, p is the poles and k is the gain.

Closed-loop system transfer function. The closed-loop system transfer function can be obtained using the *feedback* command. This command assumes by default a system a system with unity gain negative feedback. The closed-loop transfer function of the above system is thus given by

```
>> G = tf(num,den);
>> sys = feedback(G,1)
```

Transfer function:

```
        3
-----------------
s^2 + 3s + 12
```

Series and parallel connected transfer functions. Consider two serially connected transfer functions $G(s)H(s)$. The overall transfer function can be obtained as

```
series(G,H)
```

where G and H are the transfer functions $G(s)$ and $H(s)$, respectively, in MATLAB representation. For example, if

$$G(s) = \frac{1}{s^2 + 3s + 4} \quad \text{and} \quad H(s) = \frac{2}{s + 5}$$

then $G(s)H(s)$ can be obtained from

```
>> G = tf(1,[1  3  4]);
>> H = tf(2,[1  5]);
>> GH = series(G,H)
```

Transfer function:

```
              2
   -------------------------
   s^3 + 8 s^2 + 19 s + 20
```

Similarly, the *parallel* statement can be used to reduce the transfer functions connected in parallel.

Factored transfer functions. If a transfer function is in factored form, it can be entered using the *conv* command. For example, if

$$G(s) = \frac{(s + 4)}{(s + 1)(s + 2)}$$

then it can be entered into MATLAB as

```
>> num = [1  4];
>> den1 = [1  1];
>> den2 = [1  2];
>> den = conv(den1,den2);
```

Similarly, for the transfer function

$$G(s) = \frac{2}{(s + 1)(s + 2)(s + 4)}$$

we can write

```
>> num = 2;
>> den1 = [1  1];
>> den2 = [1  2];
>> den3 = [1  4];
>> den = conv(den1,conv(den2,den3));
```

Inverse Laplace transforms. The MATLAB command *residue* is used to obtain the inverse Laplace transform of a transfer function by finding the coefficients of the partial fraction expansion. The partial fraction expansion is assumed to be in the following format:

$$Y(s) = \frac{r(1)}{s - p(1)} + \frac{r(2)}{s - p(2)} + \frac{r(3)}{s - p(3)} + \cdots + \frac{r(n)}{s - p(n)} + k(s).$$

As an example, suppose that we wish to find the coefficients A, B and C of the partial fraction expansion

$$Y(s) = \frac{1}{(s+1)(s+2)(s+3)} = \frac{A}{s+1} + \frac{B}{s+2} + \frac{C}{s+3}.$$

The required MATLAB commands are

```
>> num = [1];
>> den = conv([1 1],conv([1 2],[1 3]));
>> [r,p,k] = residue(num,den)
```

r =

```
    0.5000
   -1.0000
    0.5000
```

p =

```
    3.0000
   -2.0000
   -1.0000
```

k =

```
    [ ]
```

The required partial fraction expansion is then

$$Y(s) = \frac{0.5}{s+1} - \frac{1}{s+2} + \frac{0.5}{s+3}.$$

B.2.2 Discrete-Time Systems

The Control System Toolbox also supports the design and analysis of discrete-time systems. Some of the most commonly used discrete-time system commands and algorithms are given in this section.

Discretizing a continuous transfer function. The C2d function can be used to discretize a continuous system transfer function. The sampling time must be specified. The default method of discretization is zero-order hold at the inputs, but other methods such as linear interpolation or bilinear approximation can be selected. For example, consider the continuous-time system transfer function

$$G(s) = \frac{1}{s+4}.$$

Assuming the sampling period is 0.1 s we can convert the transfer function to discrete time using the following commands:

```
>> G = tf(1, [1,4]);
>> Gz = c2d(G, 0.1)
```

Transfer function:

```
  0.08242
-----------
z - 0.6703
```

Sampling time: 0.1

Thus, the required discrete time transfer function is

$$G(z) = \frac{0.08242}{z - 0.6703}.$$

In the following example we convert a second-order continuous-time system,

$$G(s) = \frac{4}{s^2 + 4s + 2},$$

to discrete form, with sampling time 1 s:

```
>> G = tf(4,[1 4 2]);
>> Gz = c2d(G, 1)
```

Transfer function:

```
   0.6697 z + 0.1878
---------------------------
z^2 - 0.5896 z + 0.01832
```

Sampling time: 1

Poles and zeros. The poles and zeros can be obtained as follows:

```
>> [z,p,k] = zpkdata(Gz,'v')

z =
   -0.2804

p =
    0.5567
    0.0329

k =
    0.6697
```

Thus, $G(z)$ has one zero at -0.2804 and two poles at 0.5567 and 0.0329. The d.c. gain is 0.6697.

The positions of the poles and zeros can be plotted on the complex plane using the command

```
>> pzmap(num,den)
```

Also, the positions of the poles and zeros and the unit circle in the z-plane can be plotted using the command

```
>> zplane(num,den)
```

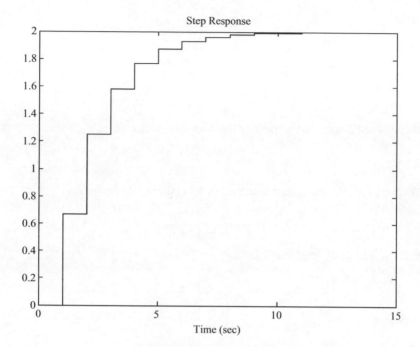

Figure B.8   Step response

Step response. The unit step response of $G(z)$ is obtained from

```
>> num = [0  0.6697  0.1878];
>> den = [1  -0.5896  0.01832];
>> dstep(num,den)
```

and the response obtained is shown in Figure B.8.

Impulse response. The impulse response of $G(z)$ is obtained by writing

```
>> num = [0  0.6697  0.1878];
>> den = [1  -0.5896  0.01832];
>> dimpulse(num,den)
```

and the response is shown in Figure B.9.

Root locus. The root locus diagram with lines of constant damping factor and lines of constant natural frequency is shown in Figure B.10 and is obtained from

```
>> zgrid('new');
>> rlocus(num,den)
```

The gain and the roots at any point on the locus can interactively be found using the command

```
>> [k,p] = rlocfind(num,den)
```

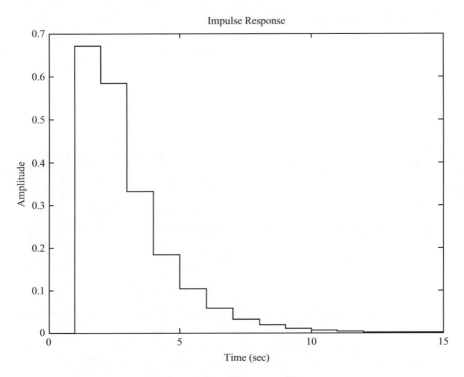

Figure B.9 Impulse response

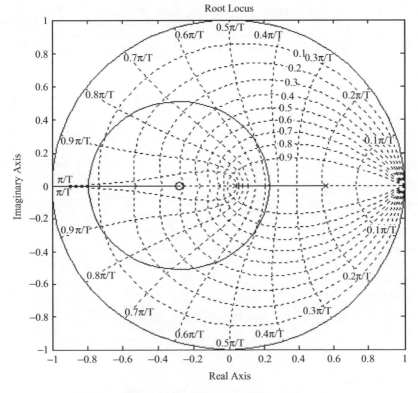

Figure B.10 Root locus diagram

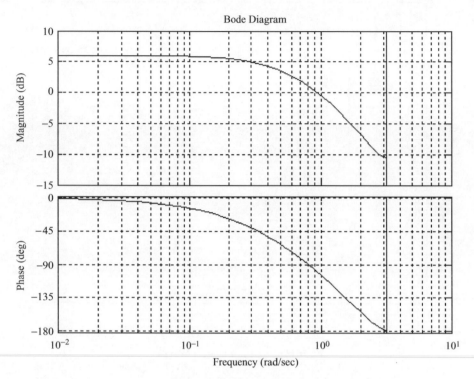

Figure B.11 Bode diagram

Bode diagram. The Bode diagram of a discrete time system can be obtained (assuming a sampling time of 1 s) as

```
>> dbode(num,den,1);
>> grid
```

The graph obtained is shown in Figure B.11.

Nyquist diagram. The Nyquist diagram of a discrete time system can be obtained as (assuming a sampling time of 1 s)

```
>> dnyquist(num,den,1);
```

The graph obtained is shown in Figure B.12.

z-Transform. The z-transform of a time function can be found using the MATLAB function *ztrans*. Some examples are given below.
    The z-transform of $f(kT) = kT$ is found as

```
>> syms k T;
>> ztrans(k*T)

ans =
    T*z/(z-1)^2
```

Nyquist Diagram

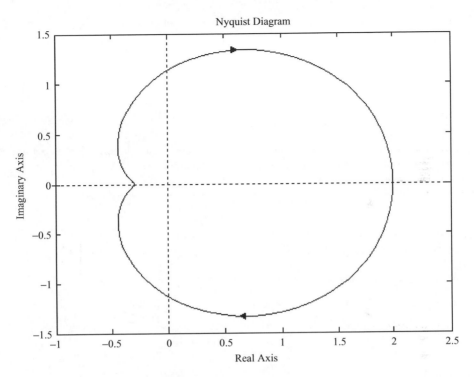

Figure B.12   Nyquist diagram

Notice that k and T are defined as symbols.

Similarly, the z-transform of $f(kT) = \sin(akT)$ is found as follows:

```
>> syms a k T;
>> f = sin(a*k*T);
>> ztrans(f)

ans =

    z*sin(a*T)/(z^2-2*z*cos(a*T)+1)
```

or

```
>> pretty(ans)

z sin(a T)
---------------------
 2
z - 2 z cos(a T) + 1
```

Inverse z-transform. The inverse z-transform of a function can be found using the *iztrans* function. Some examples are given below.

The inverse z-transform of $F(z) = Tz/(z-1)^2$ is obtained as follows:

```
>> f = T*z/(z-1)^2;
>> iztrans(f)

ans =
     T*n
```

Notice that the default independent variable is n.

Coefficients of partial fraction expansion. MATLAB can be used to determine the coefficients of the partial fraction expansion. Some examples are given below.

Consider the transfer function

$$G(z) = \frac{2z^2 - z}{z^2 - 3z + 2}.$$

We usually expand the term $G(z)/z$ which gives a form which is usually easy to look up in the inverse z-transform tables. Thus,

$$\frac{G(z)}{z} = \frac{2z - 1}{z^2 - 3z + 2}.$$

The coefficients of the partial fraction expansion are found as follows:

```
>> [r,p,k] = residue([2 -1], [1 -3 2])
```

```
r =
     3
    -1

p =
     2
     1

k =
     []
```

where r are the residues, p are the poles and k are the direct terms. Thus,

$$\frac{G(z)}{z} = \frac{3}{z-2} - \frac{1}{z-1}$$

and

$$G(z) = \frac{3z}{z-2} - \frac{z}{z-1}$$

The time function can easily be found using z-transform tables.

Another example is given below where there is one direct term. Consider the transfer function

$$\frac{G(z)}{z} = \frac{2z^2 + 2z - 1}{z^2 - 3z + 2}.$$

The coefficients are found from

```
>> [r,p,k] = residue([2 2 -1], [1 -3 2])
```

```
r =
    11
    -3
```

```
p =
     2
     1
```

```
k =
     2
```

Thus,

$$\frac{G(z)}{z} = \frac{11}{z-2} - \frac{3}{z-1} + 2$$

or

$$G(z) = \frac{11z}{z-2} - \frac{3z}{z-1} + 2z$$

and the inverse z-transform can be found using z-transform tables.

The following example has a double pole. Consider the transfer function

$$\frac{G(z)}{z} = \frac{z^2 + 4z - 1}{z^3 - 5z^2 + 8z - 4}.$$

The coefficients are found from

```
>> [r,p,k] = residue([0  1  4  -1], [1  -5  8  -4])
```

```
r =
    -3.0000
    11.0000
     4.0000
```

```
p =
     2.0000
     2.0000
     1.0000
```

```
k =
     [ ]
```

There are two poles at $z = 2$, and this implies that there is a double root. The first residue is for the first-order term for the double root, and the second residue is for the second-order term for the double root. Thus,

$$\frac{G(z)}{z} = \frac{-3}{z-2} + \frac{11}{(z-2)^2} + \frac{4}{z-1}$$

or

$$G(z) = -\frac{3z}{z-2} + \frac{11z}{(z-2)^2} + \frac{4z}{z-1}.$$

The MATLAB command *residuez* can be used to compute the partial fraction expansion when the transfer function is written in powers of z^{-1}.

Index

Printed and bound by CPI Group (UK) Ltd, Croydon, CR0 4YY